Current Issues in Toxicology

Current Issues in Toxicology

Sponsored by the
International Life Sciences Institute

Edited by
K.S. Khera, H.C. Grice, and D.J. Clegg

Interpretation and Extrapolation of Reproductive Data to Establish Human Safety Standards

Springer-Verlag New York Berlin Heidelberg
London Paris Tokyo

K.S. Khera
Ottawa, Ontario, Canada

H.C. Grice
Nepean, Ontario, Canada

D.J. Clegg
Ottawa, Ontario, Canada

Library of Congress Cataloging-in-Publication Data
Interpretation and extrapolation of reproductive data to establish
human safety standards/edited by K.S. Khera, H.C. Grice, and D.J. Clegg
p. c.m. – (Current issues in toxicology)
ISBN 0-387-96962-4
1. Reproductive toxicology–Technique. 2. Health risk assessment.
I. Khera, K. S. II. Grice, H. C. III. Clegg, D. J. IV. Series.
[DNLM: 1. Reproduction–drug effects. 2. Teratogens–toxicity.
3. Toxicology–methods. 4. Toxicology–standards. WG 205 I618]
RA1224.2.I57 1989
618.3–dc 19
DNLM/DLC 89-4080

Media conversion by Publishers Service, Bozeman, Montana.
Printed and bound by R.R. Donnelley & Sons, Harrisonburg, Virginia.
Printed in the United States of America.

9 8 7 6 5 4 3 2 1

ISBN 0-387-96962-4 Springer-Verlag New York Berlin Heidelberg
ISBN 3-540-96962-4 Springer-Verlag Berlin Heidelberg New York

Series Preface

The International Life Sciences Institute (ILSI) is a scientific foundation that addresses critical health and safety issues of national and international concern. ILSI promotes international cooperation by providing the mechanism for scientists from government, industry, and universities to work together on cooperative programs to generate and disseminate scientific data. The members and trustees of the institute believe that questions regarding health and safety are best resolved when scientists can examine and discuss issues as an independent body separate from the political pressures of individual countries and the economic concerns of individual companies.

Frequently, meaningful assessment of the risk of a test substance is hindered by the inherent inconsistencies in the system. The development and refinement of methods and systems to evaluate the safety of chemicals have evolved in a rapid and largely unplanned fashion.

Attempts to improve the system have mainly been directed toward broad general concerns, with little attention being given to specific problems or issues. A failure to resolve these problems has frequently resulted in increased testing costs and complications in the assessment and extrapolation of the results to humans. Publicity surrounding toxicologic issues makes it difficult for governments to deal effectively with these problems.

In response to these difficulties, ILSI has assembled highly qualified and renowned scientists from research institutes, universities, government, and industry with relevant scientific knowledge and expertise regarding the issues that complicate risk assessment procedures.

This series, *Current Issues in Toxicology*, is the result of the endeavors of these international scientists. It also exemplifies the commitment of ILSI to promote a better understanding of critical safety issues. Throughout this series, an attempt is made not only to examine the factors that influence the evaluation of the safety of chemicals but also to develop principles, recommend guidelines, and define areas requiring additional research.

Nepean, Ontario, Canada H.C. Grice

Contents

Contributors

M.S. Christian, Ph.D.
Argus Research Labs, Inc.
Horsham, Pennsylvania, U.S.A.

D. Clegg, M.Sc.
Health Protection Branch
Health and Welfare Canada
Ottawa, Ontario, Canada

D. Colin, Ph.D.
Universite du Quebec
Montreal, Quebec, Canada

K. Hemminki, M.D.
Institute of Occupational Health
Helsinki, Finland

R.M. Hoar, Ph.D.
Argus Research Labs, Inc.
Horsham, Pennsylvania, U.S.A.

K.S. Khera, D.Sc.
Health Protection Branch
Health and Welfare Canada
Ottawa, Ontario, Canada

D. Kirkpatrick, B.Sc., M.Sc.
Health Protection Branch
Health and Welfare Canada
Ottawa, Ontario, Canada

D. Krewski, Ph.D.
Environmental Health Directorate
Health and Welfare Canada
Ottawa, Ontario, Canada

R. Robertson, Ph.D.
Merck, Sharpe & Dohme Research Laboratories
West Point, Pennsylvania, U.S.A.

J. Rundell, Ph.D.
Molecular Toxicology, Inc.
College Park, Maryland, U.S.A.

G. Sherman, Ph.D.
Laboratory Centre for Disease Control
Health and Welfare Canada
Ottawa, Ontario, Canada

F.M. Sullivan, B.Sc. (Hons.)
UMDS (Guy's Hospital)
University of London
London, U.K.

T. Tanimura, M.D.
Kinki University
Sayama-cho, Osaka, Japan

T. Yanagimoto, Ph.D.
The Institute of Statistical Mathematics
Tokyo, Japan

Chapter I

Introduction

This monograph focuses on extrapolating data from animal reproduction studies to permit the assessment of exposure levels acceptable for human reproduction that are consistent with the maintenance of public health.

The interpretation and extrapolation of animal data for predicting human health effects from reproductive toxicants is a complex process with many uncertainties. This document attempts to outline some of the factors involved in the process, taking into account biological responses relevant to the test compound in experimental animals and humans. The strengths and limitations of the test systems employed also are discussed.

The adverse effects that a chemical can produce in the reproductive system are determined by an interplay of endogenous and exogenous factors. These include the dose of a chemical and its pharmacokinetic and metabolic interactions with cells and intracellular constituents. Other factors, such as genetic, endocrine, immunological, and nutritional status as well as environmental influences, also play an important role.

The key to extrapolation of adverse reproductive responses observed in animal studies to humans lies in an understanding of the mechanisms by which a test substance exerts its effects. Until these mechanisms are understood, risk estimation is largely empirical.

The reliability of controlling human exposure levels for safe human prenatal and postnatal development largely depends upon the accuracy of toxicologic assessment of reproductive data from animal testing of natural and synthetic chemicals. Variations in anatomical, physiological, and biochemical characteristics among different species may contribute to differences in response to chemical agents. Hence, sound extrapolation of findings from one species to another requires an understanding of the fundamental mechanisms underlying a response.

If it can be determined that the experimental model will reflect what is likely to occur in humans, then judgments on extrapolation are more sound and can be made with an increased degree of certainty that is reassuring. Unfortunately, data for adverse reproductive effects, on which such judgments can be based, are seldom available. Without these data, the uncertainty in the process must be bolstered with the application of judicious safety factors derived from effect and no-observed-effect levels in test animals. Once an acceptable intake has been established, it must be compared with probable human intake or exposure levels to determine if these are safe. Furthermore, for most chemicals there is usually little available historical data to permit reliance on precedent. For example, only a few agents have been established as being human teratogens on the basis of sound clinical or epidemiological data. On the other hand, many agents have been shown experimentally to induce teratogenic effects, usually at high dosages, but such effects have not been observed in humans, possibly because humans are seldom exposed to similarly high dosages. In general, the principal source of evidence for adverse reproductive potential is toxicity studies conducted in laboratory animals.

This monograph contains some departures from existing principles and practices in reproductive toxicity. For example, despite recent recognition of the role and relevance of maternal toxicity in fetal development, this has not always been considered in assessing the teratogenicity of chemicals. Lack of attention to chemically induced maternal toxicity has sometimes led to an incorrect assessment and interpretation of developmental and reproductive effects. In this monograph the relative selective action of a chemical on the fetus or the mother is considered. Chemicals are classified into categories by the nature of the events they induce in the dam or the fetus. Based on this criterion, the development of safety factors that are appropriate and relevant to the induced events is outlined.

In Chapter II the terminology and definitions used in the monograph are provided. An attempt has been made to give precise meaning to the terms. It is recognized that generally accepted terminology and definitions of key terms are prerequisites for achieving worldwide uniformity and understanding. The design, conduct, and evaluation of teratogenicity (embryotoxicity) studies is outlined. Maternal and embryofetal toxicity are considered in Chapter III. There is no attempt to provide detailed procedures to be followed in the design and conduct of studies because numerous guidelines that provide details are available. Rather, the chapter outlines the salient features that help determine if a study is suitable for evaluation purposes. The implications of maternal toxic effects for the fetus are difficult for regulatory authorities to deal with because distinct criteria to reveal the existence and significance of these effects have not been completely worked out. The use of structure-activity relationships and control data in the process is described in Chapter III, with the recognition that knowledge of chemical structure and its relevance to teratogenic activity are limited.

In Chapter IV the design, conduct, and evaluation of multigeneration studies are considered. Multigeneration studies (being in essence screening studies) cannot be considered in isolation but need to be considered in conjunction with the overall database relevant to the toxicity of the test material.

Behavioral-functional teratology is considered in Chapter V. Methods currently in use or under development are outlined, including CNS effects, functional testing procedures, and current status of testing procedures and their uses. Two concepts concerning the purpose of behavioral-functional studies are evident. If the purpose is to study mechanisms and developmental changes, all organ systems are subject to investigation, and the most sensitive organ system in which alterations occur should be identified as the target. If the purpose is to determine the type of effect and the no-effect level for risk assessment, the most sensitive relative variable in the adult or offspring must be considered.

In Chapter VI a variety of in vitro screening procedures is considered. These procedures include assay systems using mammalian species, submammalian vertebrate species, and invertebrate models. The principal role of currently available in vitro tests is to examine mechanisms of developmental toxicity. At present these systems do not provide a basis for human risk assessment.

In Chapter VII statistical methods for the analysis of data derived from reproductive toxicity studies are discussed. In particular, statistical tests are described that may be used to compare the rate of occurrence of terata and other reproductive anomalies in animals exposed to the test compounds and in unexposed controls. These procedures may be used to establish a no-observed-effect level (NOEL) at which the rate of occurrence of adverse effects is not significantly increased relative to the background rate observed in controls. In addition, statistical models that may be used to describe the dose-response relationship for such effects are discussed. These models provide a quantitative description of the manner in which the response rate varies with dose and may be useful in explaining potential risks at lower doses. Both the statistical tests for teratogenic and reproductive effects and the dose-response models treat the litter as the experimental unit, thereby allowing for the presence of litter effects. Although this chapter is somewhat more technical than previous chapters, it will be of interest to statistical scientists collaborating in the analysis and interpretation of reproductive toxicity studies.

In Chapter VIII the conduct and characteristics of acceptable epidemiological studies on human reproduction are outlined. The increased participation of women in the labor force has been mirrored by a trend for women to continue working for a greater proportion of pregnancy. Despite this trend, very few reliable epidemiological data exist regarding health effects on the mother and/or fetus of specific chemicals encountered in the workplace.

In Chapter IX the assessment of human risk from reproductive data is considered. Three steps in the process are outlined: the conduct of studies in laboratory animals and humans, the evaluation of these data, and the extrapolation of human risk from the data. The judicious extrapolation of human risk from toxicity studies requires a broad database, a broad base of knowledge, and familiarity with the performance of complex animal studies. The basic criteria used to evaluate fetal malformations and other embryofetal toxic effects are identical for all chemicals. However, the weighting of these criteria is frequently modified by ancillary data (e.g., pharmacokinetics). A basic problem with the system is an inability to utilize underlying mechanisms of action because of a lack of knowledge of such mechanisms for individual chemicals. In the case of behavioral and functional studies, the comparative toxicity in the mother and offspring must be considered. Subtle changes in compromised offspring should not be given undue importance if the parents are also compromised. The principal role of currently available in vitro cytotoxicity tests has been the prescreening of chemicals to distinguish those with similar activity. These tests have not yet been used in risk assessment.

The risk assessment from epidemiological studies involves the use of operational criteria that can help determine the likelihood of a causal relationship between an exposure and a reproductive effect. Key aspects of effective environmental health programs are monitoring and assessing adverse human health effects from exposure to environmental agents to confirm that regulatory decision has been correct and that control measures are adequate.

In Chapter X the guidelines for reproductive toxicity tests in different countries are summarized and a proposal for international harmonization is presented. The regulatory judgment on risk assessment of reproductive effects may vary from country to country. Although the judgment may be influenced by political and economic needs, the scientific assessment should reflect a high degree of international agreement, especially when a good database is available. It is suggested that many of the variations in regulatory guidelines may not be of sufficient scientific importance to require the conduct of several similar studies. The time is right and the need is great for scientific organizations, such as the International Federation of Teratology Societies in collaboration with the World Health Organization and other international bodies, to begin to achieve international agreement on harmonized guidelines.

Chapter II

Terminology and Definitions

In this chapter terminology and definitions are provided to give precise meaning to the terms that are used in this monograph. The use and meaning of terms employed in teratology studies vary between laboratories (Black and Marks, 1986) and regulatory agencies of different countries. An attempt to achieve worldwide uniformity in the assessment of human risks, particularly those involving prenatal and postnatal development, is an absolute need in view of rising international trade in foods, drugs, chemicals, and other regulated consumer products. Generally accepted terminology and definitions of key terms are necessary prerequisites for achieving such uniformity.

Maternal toxicity is a transitory or permanent pathologic state of health or alteration in maternal physiology and/or behavior with the potential to cause adverse effects in the offspring during embryofetal or postnatal development.

A *malformation* is a serious anatomical defect of prenatal origin that may or may not be compatible with survival. If compatible, it will adversely influence development, growth, morphology, physiological functions, fertility, and/or longevity.

An *abnormality* is an abnormal change in structure, appearance, function, or behavior. As distinct from a malformation, an abnormality does not influence postnatal development.

An *aberration* is a minor structural change. It may be a retardation (a provisional delay in morphogenesis), a variation (external appearance controlled by genetic and extragenetic factors), or a deviation (resulting from altered differentiation) (for details, see Khera, 1981).

A *teratogen* is an intrinsic or extrinsic factor or an alteration in maternal homeostasis that induces, during prenatal development, a permanent structural or functional abnormality in the fetus, which is detected in a prenatal or postnatal examination (Clegg, 1971). If malformations are induced at apparently nonmaternotoxic doses, the agent is a true, or selective, teratogen.

Embryotoxicity signifies embryonic loss during the early preimplantation or postimplantation stages of pregnancy.

Fetotoxicity is any prenatally initiated toxic manifestation observed in a fetus (death, body-weight reduction, delayed ossification, or functional defect), which may or may not be related to toxic effects in the mother and which although initiated in utero appears during prenatal or postnatal development. Any of these effects resulting from the direct action of a test agent on the embryo or fetus (as defined in *Dorland's Medical Dictionary*) and occurring at doses far below those toxic for the mother should be regarded as suggestive of true, or selective, embryo- or fetotoxicity.

The term *embryo-fetus* is used when the stage of prenatal development at the time of initial insult is not known.

Developmental toxicity represents any adverse effect on development (morphologic, physiologic, or functional) initiated prenatally and appearing during the lifetime of the progeny.

Reproductive toxicity deals with toxic effects on any aspect of reproduction of offspring proceeding from the development of gametes and their fusion.

A *threshold level* is the highest level of a chemical or test substance, obtained by the best possible estimates from experimental data, that is judged insufficient to produce an adverse effect on prenatal and postnatal development in humans or animals.

A *safety factor* is an arbitrarily determined numerical value that allows for the uncertainty in estimating potential human effects from animal data and for unforeseen individual sensitivity in the human population. An acceptable dose is derived from the maximum dose at which no adverse effects are observed divided by a safety factor (WHO, 1958).

Reproductive hazard is the likelihood that a chemical or test substance will cause an adverse effect, including gross anatomical defects, on any aspect of reproduction under the conditions of its production or use (modified from WHO, 1978).

Risk is the expected frequency of undesirable effects arising from a given level of exposure to a chemical or test substance. An acceptable risk is based on the probability of occurrence of adverse effects and their consequences on human health compared with the benefits from the use of a substance.

Safety is the practical certainty that an adverse effect will not result when a substance is used in the proposed quantity or manner.

References

Black, D.L. and Marks, T.A. (1986) Inconsistent use of terminology in animal developmental toxicology studies: A discussion. *Teratology, 33*, 333–338.

Clegg, D.J. (1971) Teratology. *Annu. Rev. Pharmacol.* 11:409–424.

Dorland's Medical Dictionary (1965) 24th ed., pp. 478–547, W.B. Saunders, Philadelphia.

Khera, K.S. (1981) Common fetal aberrations and their teratologic significance: A review. *Fund. Appl. Toxicol.* 1:13–18.

WHO (World Health Organization) (1958) *Methods for Toxicologic Testing of Food Additives*, Technical Report Series, No. 144, pp. 1–19, World Health Organization, Geneva.

WHO (World Health Organization) (1978) *Principles and Methods for Evaluating the Toxicity of Chemicals*, Part I, Environmental Health Criteria 6, p. 19, World Health Organization, Geneva.

Chapter III

Teratogenicity (Embryotoxicity) Studies: Design, Conduct, and Evaluation

3.1. Introduction

The principal source of evidence for adverse effects on fetal development that are recognized at term is teratogenicity studies in experimental animals. The design of these studies should be appropriate for the intended purpose, and their conduct should conform to high standards. This is necessary to provide assurance that the data used in risk estimation are sound. In this chapter only basic principles related to these elements are discussed. References cited in the text should be consulted for details regarding the design and conduct of such studies.

The aim of a teratogenicity, or prenatal, study is to investigate the potential of a test agent to cause fetal malformations, embryo-fetal death, fetal weight (or length) reduction, or adverse effects on maternal health. A number of guidelines are available for the design and conduct of studies appropriate for subsequent risk assessment (FDA, 1970; NHW, 1973; NAS, 1977; EPA, 1981, 1984a; OECD, 1981; CMEA, 1982; WHO, 1984). A teratogenicity study is usually conducted in at least two stages—primary and/or secondary. At the primary stage, the lowest doses that cause embryo-fetal effects and/or overt maternal toxicity as well as the type of embryo-fetal effects and toxicity produced are determined. Secondary, or special, studies are designed to determine the specificity of malformations, mechanisms of action, target cells, distribution and metabolism, and fetal levels of the test chemical or its metabolites. These studies are helpful in defining specific human risks from animal data.

The reliability of estimated teratogenic risk depends upon well-conducted experiments with adequate numbers of test animals, good animal husbandry practices, adequate maternal and fetal appraisals, and the availability of accurate experimental data. Information of questionable quality obtained from inadequately conducted experiments should not be used for risk analysis. The design of animal studies is appropriate for consideration of human-risk estimation when the following are clearly demonstrated: (1) presence or absence of maternal toxicity at each dose tested, (2) dose regimen and route of exposure similar to the major human exposure condition, (3) types of adverse fetal effects, and (4) doses showing positive and no observable effects on fetal development or maternal health together with the shape of dose-response curves.

3.2. Protocol

The protocol clearly states the objective of the study and provides details on the test chemical (code number, purity, composition, vehicle for suspension, and solubility), test species (descriptions of strain, body-weight range, age, source of supply, and method of animal identification), and experimental design (doses, route and duration of treatment, identification of and methods for measuring fetal and maternal endpoints, and intended methods of statistical analysis). Any deviations from the protocol during the conduct of the study must be justified and recorded as amendments.

3.3. Conduct of the Study

The test chemical is of defined purity; its stability, homogeneity, and concentration in the defined vehicle or diet for the duration of the dosing period is supported by analyses of an adequate number of samples. Exposure conditions simulate the major human-exposure conditions with respect to the route and duration. A suitable selection is made of dose levels – high (maximum maternally or fetally tolerated), low (no adverse fetal effects), and medium (some effects on fetuses or maternal animals). The mated females are dosed daily during the period of organogenesis, from days 6–16 of pregnancy in rats, days 6–15 in mice, and days 6–19 in rabbits, with the day of insemination considered as day 0 of pregnancy. Experimental animals are preferably from an outbred, healthy stock that is sensitive to known teratogenic agents. Animal husbandry in the animal colony includes efficient quarantine facilities, effective disease control, and stringent restrictions on intercurrent contagious diseases. During illness, information is recorded on the diagnosis, medication, duration of treatment, and rooms and site where animals are retained. All test animals are appropriately identified and kept in clean, hygienic, and controlled housing conditions. Raw data should contain detailed records of all findings. Useful guidance on good laboratory practice applicable to planning and conducting studies and recording data is available (FDA, 1978; EPA, 1983a, 1983b; ECETOC, 1979).

Maternal effects that should be recorded are any disease process, including intercurrent disease; toxic effects; pathological gross and microscopic changes; treatment-related pharmacologic and toxicologic signs in behavior; abortion; death; and increase or decrease in body weight or consumption of feed and water. Effects in the conceptus that should be recorded are resorption, fetal death, live fetuses, mean body weight of male and female fetuses (separate and combined), malformed fetuses, sex ratio, type and dose-relatedness of individual fetal malformations, and aberrations; the location of the fetus in utero should also be noted. Fetuses for visceral and skeletal examination are randomly selected and examined blind.

3.4. Evaluation of Animal Data

3.4.1. Statistical Analysis

For the evaluation of chemically induced prenatal effects, data are analyzed separately with the fetus and litter as separate experimental units or together by combining the fetal and litter data. Each fetal effect (resorption, fetal death, fetal weight or length reduction, and individual malformation) is analyzed independently for a dose-response relationship, although it is realized that the dose range for each individual effect may be restricted. The above fetal effects may influence or obscure each other's appearance, even though the underlying mechanisms for their manifestation do not seem to be similar. For the purpose of extrapolation, a particular adverse effect that is observed at the lowest dose and its dose-effect curve usually carry the highest significance.

3.4.2. Maternal Toxicity

The concept of maternal toxicity was first described by Chaube and Murphy (1968), who compared LD_{50} values for the mother and fetus for several chemicals. A number of workers have grappled with the complexity of evaluating maternal toxicity and its effect on fetal development. According to Johnson

Table 3-1. Signs of Overt Toxicity in the Maternal Organism

1. Daily body weight changes and effects on foods and water consumption after initiation of dosing.
2. Time to onset of signs; duration and type of signs.
3. Apparent changes in respiration, alertness, posture, movement within the cage, respiration, color of mucous membranes, behavior (aggressive, depressed, lethargic, sedated), hair and coat appearance, color of urine, and frequency of urination.
4. Other signs such as nasal discharge, diarrhea, salivation, tremor, convulsions, and coma.
5. Death and necropsy findings.

(1981) the hazard potential of a chemical is identified by an index obtained from the adult LD_{50} dose divided by the dose that is toxic for development. A high index for a chemical disrupting development at doses that are not toxic to adults indicates a potential teratogenic hazard, whereas a low index (<0.1) does not. Subsequently, Johnson suggested the use of A:D ratio obtained from a dose toxic in the mother (A) divided by a dose toxic for development (D), as determined from Segment 2 studies (Johnson, 1984). Fabro et al. (1982) computed a relative teratogenic index from the minimum adult lethal dose of LD_{01} divided by the minimum teratogenic dose (tD_{05}, a dose causing malformations in 5% of live fetuses). This index and tD_{05} were the respective quantitative measurements of teratogenic hazard and teratogenic potency for the species tested. The study by Clark et al. (1984) has demonstrated the importance of assessing maternal toxicity for its effects on fetal development.

To assess the existence of toxicity in the maternal organism, a number of maternal endpoints have been suggested (EPA, 1984b). Several variables indicative of toxicity in the gravida are suggested here (Table 3-1), but the choice of a scientifically valid yet simple method is by no means easy or obvious. Perhaps a simple method for the assessment of this complicated phenomenon will emerge when more data become available. The determination of toxicity during pregnancy needs a shift in emphasis toward judging its absence rather than presence, because the major concern is the implication of maternal disease for the fetus and much less the maternal disease per se. Death (and its incidence at different dosages of a toxicant) is a definite response, but the associated fetal implications seem to depend more upon the duration and intensity of maternal disease preceding death than on the death alone. A number of potent poisons may cause instant maternal death and consequent fetal death, with little influence on embryo-fetal development.

Any attempt to relate fetal development to a chemical's adult lethal dose (LD_{50}, LD_{05}, etc.) or to signs of toxicity in a nonpregnant animal ignores the basic fact that the stress of pregnancy may render a pregnant animal more sensitive than a nonpregnant counterpart to the toxic effects of a chemical. However, the converse may occur, e.g., the situation of decreased maternal toxicity with some heavy metals during pregnancy. LD_{50} values in pregnant animals to be meaningful in teratogenicity studies, should provide additional information on (1) the time-interval between dosing and death and between the first appearance of toxicity signs and death of test animals and (2) the signs of toxicity. Further, to demonstrate a maternal-fetal relationship, the day or stage of pregnancy when fetal development is disrupted and maternotoxic signs occurred should be related.

The implications of maternal toxic effects for the fetus are more difficult for regulatory authorities to cope with because distinct criteria to reveal the existence and significance of these effects have not been completely worked out. Requiring data on too many endpoints with no established validity would lead to a situation of preferential selection of endpoints by an investigator wishing to demonstrate the existence or the absence of toxicity. The intensity and duration of signs (Table 3-1) immediately after dosing with the test chemical are probably among the most reliable criteria of maternotoxic effects. Although extremely toxic effects detected from clinical signs are quite obvious, at present it is not possible to draw a line separating low levels of overt changes in behavior from covert homeostatic alterations that are not manifested as external signs. Reduced maternal weight gain during pregnancy and other maternal disorders have been associated with excessive prenatal mortality in humans (Naeye, 1980) and could also be useful variables in animal studies. Before an early resorption of its whole litter, a pregnant animal would gain body weight at a rate similar to that of a nonpregnant animal. An adjustment for the number of viable fetuses and their reduced fetal weight could result in overcorrection. In other words, a pregnant and a nonpregnant female (which has resorbed all embryos) may not gain body weight at the same rate.

Although not essential, it would be highly desirable to define maternal changes in hematology, clinical chemistry, urinalysis, and histology of organ systems (see methods in Page et al., 1980).

3.4.3. Embryo-Fetal Toxicity

3.4.3.1. Aberrations and Malformations. The ascertainment of malformations and aberrations depends on the vigor with which the search is pursued and the competence, experience, and patience of the observer. These factors vary from study to study and determine to a large extent the accuracy of estimating anomalies. Underestimation may occur because a number of fetuses or offspring may carry subtle or occult anomalies (e.g., transposition of great arteries) that may not be noticed in a prenatal or postnatal study because a thorough teratologic examination of offspring is not conducted unless indicated by clinical signs. Interobserver bias may, at times, make it difficult to obtain complete and consistent diagnostic accuracy between laboratories.

Structural aberrations including transitory developmental delays, such as retarded ossification of sternal and appendicular bones, nonsymptomatic permanent alterations (including supernumerary ribs), and anatomical changes that are repairable during postnatal development (e.g., wavy ribs), have never been truly defined for their teratologic significance (Khera, 1981). The presence of supernumerary ribs has been related to nonspecific maternal toxicity in the mouse (Khera, 1984; Kavlock et al., 1985). The frequency of supernumerary ribs in pups from control groups and groups prenatally treated with aspirin has been shown to decline to essentially zero (Wickramaratne et al., 1987). The significance of wavy ribs during postnatal development in teratologic evaluation of the fetal rat is less certain (Sterz et al., 1985). Limbs with abnormally increased flexion of joints without tissue deformity have frequently been interpreted as malformations. The joints in all such cases should be manipulated in an attempt to determine whether the flexion is spurious, temporary, or permanent (Pinto-Machado, 1985). The biological cause of a number of malformations occurring primarily at low incidence at maternotoxic doses with a rather obscure dose-response relationship is not clear. A tentative list of malformations associated with maternal toxicity has been reported (Table 3-2; Khera, 1985). It is emphasized that any of these structural aberrations or malformations could also result from a chemical's direct action in fetal tissues, independent of the toxic effects in the mother.

All major malformations are, at present, considered equally important in evaluating the teratogenic potential of chemicals. A case may be made that the significance of a malformation should be based on its compatibility with survival during the rigors of birth and prenatal and postnatal development. A chemical would more appropriately qualify as embryocidal or fetocidal, rather than teratogenic, if its exposure causes embryonic or fetal death, which is invariably preceded by a malformative change. These types of malformations are, therefore, inextricably connected with embryo-fetal death.

In humans, anencephaly, cebocephaly, hydrocephaly, myeloschisis, and cyclopia have been reported to occur at high incidences in stillborn (Richards, 1973) and spontaneously aborted embryos (Creasy and Alberman, 1976). The incidences of these malformations, depending on the type of malformation, ranged from 2.8–7.3% in 8-week-old embryos compared with 0.17–0.66% in infants (Nishimura, 1969). Anencephaly has been associated with embryo-fetal death (Book and Rayner, 1950). A frequent association between fetal and neonatal death and these CNS anomalies raises the possibility that these anomalies may be incompatible with postnatal life. A similar incompatibility between life and CNS anomalies seems to exist in species of rodents used in toxicity studies. In test animals, exencephaly, encephalocele, and spina bifida are fatal at or immediately after birth, whereas microencephaly, microceph-

Table 3-2. Fetal Malformations Tentatively Associated with Maternal Toxicity of Test Agents in Rabbit and Rodent Species

Species (day of pregnancy*)	Malformations: Frequently reported	Malformations: Less frequently reported
Hamster (8–9)	Fused ribs, exencephaly encephalocele (cranial blister), microphthalmia, or anophthalmia	Omphalocele
(10–11)		Shortened or crooked tail, cleft lip, ectrodactyly, or syndactyly
Mouse (8–10); Rat and rabbit (9, or 6–18)	Ribs: fused, extra, missing, or wavy; vertebrae: fused, retarded, missing, or split; sternebrae: missing, fused, or nonaligned	Mouse: Exencephaly, open eyes, ectrodactyly or syndactyly Rat and rabbit: Exophthalmia, anophthalmia, or microphthalmia; cleft palate; umbilical hernia; shortening or bending of long limb bones; ectrodactyly or syndactyly; crooked, short, or no tail

*The day during pregnancy when maternal toxicity occurred (day of finding sperm in vaginal smears is day 1 of pregnancy).

aly, and agenesis of the corpus callosum have postnatal survival rates ranging from 3–19½ months (Cowen and Geller, 1960). Reduced brain size has no apparent adverse effect on longevity (Coppenger and Brown, 1965). Reviews of human data published by Butler and Alberman (1969) and Mitchell et al. (1971) reveal that most cardiac anomalies are compatible with prenatal and postnatal survival. Studies are needed to define the length of survival (life tables) for individual malformations or simultaneously occurring multiple malformations to appropriately assess the teratogenic potential of chemicals.

3.4.3.2. Mortality and Weight Reduction. Preimplantation loss, resorption, late fetal death, and reduction in fetal weight are all relevant to human risk estimation. Note that there are inaccuracies inherent in the evaluation of preimplantation loss that may be a reflection of expected regression of corpora lutea when extensive resorptions occur. Inaccuracy in the estimation of implantation loss may be even greater when artificial insemination is used. Apparent preimplantation loss may also occur from infertile ova and may not be entirely limited to the loss of fertilized ova. Thus, the preimplantation loss may not be related justifiably to the exposure of a test chemical. Accurate determination of the time of conception may clarify whether the compound was first administered before or after implantation. However, frequently implantation occurs before the initial administration of test compound when a conventional teratology study protocol is used. A fetal effect that occurs at a minimum effective dose and preferably in the absence of maternal toxicity should be considered a specific fetal effect of a chemical. With a progressive increase in dose levels, an increase in the type and severity of fetal effects becomes apparent. Very high doses, which are usually but not invariably complicated by maternal toxicity, may result in multiple yet nonspecific developmental effects.

Embryo-fetal death, fetal body weight reduction, and malformation may occur independently of each other. Dimethadion (Buttar et al., 1978) and dexamethasone (Mosier et al., 1982) caused reduced fetal weight in the absence of significantly increased embryo-fetal deaths, whereas methylmercuric chloride (Fuyuta et al., 1978) and vitamin A palmitate (Hayes et al., 1981) caused resorptions unassociated with significantly reduced fetal weight. Although runts (dwarfed animals with body weights outside the range of intralitter variation) are frequently observed in control as will as in test groups, it is not known if their incidence is influenced by chemical treatment or if the runts have an increased incidence of malformations. There is an association between the development of a runt and its position at the ovarian end of the uterine horn; a runt may also arise from the fusion of two live embryos (McLaren and Michie, 1960).

3.4.3.3. Prioritization of Chemicals for Testing. The knowledge of the relationship between chemical structure and teratogenic activity has been considered for prioritizing chemical for whole-animal studies. An analysis of such relations has not been very fruitful (Schardein, 1983). The information that is available does not support the concept of estimating the safety of untested chemicals from structurally related chemicals, even though the studies reported were well designed and properly conducted. However, a structure-activity model, based on ranking probabilities of teratogenic potential with structural and physical criteria of chemicals, has been suggested for setting testing priorities of untested chemicals (Enslein et al., 1983). This approach should be assessed for practical usefulness. In combination with structural data, a model based on rank ordering of doses of chemicals causing a predetermined severity of maternal toxicity may also be promising for prioritizing chemicals for subsequent testing in animals. In a survey of studies with well-documented information on maternal toxicity, 69% (133 of 192) of the studies on chemicals revealed that malformations or resorptions occurred at doses close to or within the maternotoxic dose range (Khera, 1985).

Johnson (1987) has suggested a three-tier system to prioritize chemicals for a subsequent full-scale embryo-fetoxicity study in whole animals (Segment 2). To determine the rank order of a chemical for testing priority, the following criteria are suggested for evaluation: (1) magnitude of human exposure (level, duration, or both), (2) relative ratio of adult toxicity to developmental toxicity in an in vitro hydra bioassy (A:D ratio), and (3) results of an abbreviated test at four or more doses with eight pregnant animals at each dose following a Segment 2 protocol.

3.4.3.4. Concurrent and Historical Controls. No matter how appropriately randomization procedures are employed, it is difficult to be certain that all variations and spontaneous malformations have been randomly distributed among the experimental groups. Historical data may prove useful in deciding whether a malformation not observed or observed at a low incidence in the concurrent control group resulted by chance or was compound related.

A Task Force of Past Presidents of the Society of Toxicology (1982) has outlined propositions that may be taken as scientifically useful in the evaluation of a chemical carcinogenic response with distinctions drawn between the use of concurrent control and historical control data. These propositions could have application in considering control data in reproductive toxicology. In addition care should be taken to consider seasonal variations in malformation incidences, as well as the usual problems of intralaboratory and interlaboratory variation.

References

Book, J.A. and Rayner, S.A. (1950) A clinical and genetical study of anencephaly. *Hum. Genet.* 2:61–84.

Butler, N.R. and Alberman, E.D. (1969) *Perinatal Problems: The Second Report of the 1958 British Perinatal Mortality Survey*, p. 287, Williams & Wilkins, Baltimore.

Buttar, H.S., Dupuis, I., and Khera, K.S. (1978) Dimethadione-induced fetotoxicity in rats. *Toxicology* 9:155–164.

Chaube, S. and Murphy, M.L. (1968) The tertatogenic effects of the recent drugs active in cancer chemotherapy. *Adv. Tertol.* 3:181–237.

Clark, R.L., Robertson, R.T., Minsker, D.H., Cohen, S.M., Tocco, D.J., Allen, H.L., James, M.L., and Bokelman, D.L. (1984) Diflunisal-induced hemolytic anemia as a cause of teratogenicity in rabbits. *Teratology* 30:319–332.

CMEA (1982) Methodological recommendations, protocols, experimental studies for establishment of threshold levels of the effect of industrial compounds on reproductive function, pp. 58–75. In *Problemi promislenoi toksikologii, SEV, Postoiannaja komissija po sotrudnicestvu v oblasti zdravoohranenijs*, Moskva.

Coppenger, C.J. and Brown, S.O. (1965) Postnatal manifestations in albino rats continuously irridated during prenatal development. *Tex. Rep. Biol. Med.* 23:45–55.

Cowen, D. and Geller, M. (1960) Long-term pathological effects of prenatal X-irradiation on the central nervous system of the rat. *J. Neuropath. Exp. Neurol.* 19:488–527.

Creasy, M.R. and Alberman, E.D. (1976) Congenital malformations of the central nervous system in spontaneous abortions. *J. Med. Genet.* 13:9–16.

ECETOC (European Chemical Industry Ecology and Toxicology Centre) (1979) *Good Laboratory Practice*, Monograph No. 1, European Chemical Industry Ecology and Toxicology Centre, Brussels.

Enslein, K., Lander, T.R., and Strange, J.R. (1983) Teratogenesis: A statistical structure-activity model. *Teratogenesis Carcinog. Mutagen.* 3:289–309.

EPA (Environmental Protection Agency) (1981) *Assessment of Risks to Human Reproduction and Development of the Human Conceptus from Exposure to Environmental Substances*, Contract No. W-7405-eng-26.

EPA (Environmental Protection Agency) (1983a) Toxic substances control; good laboratory practice standards; final rule. *Fed. Regist.* 48:53922–53944.

EPA (Environmental Protection Agency) (1983b) Pesticide programs; good laboratory practice standards; final rule. *Fed. Regist.* 48:53946–53969.

EPA (Environmental Protection Agency) (1984a) *Risk Assessment and Management: Framework for Decision Making*, EPA 600/9-85-002.

EPA (Environmental Protection Agency) (1984b) Proposed guidelines for the health assessment of suspect developmental toxicants. *Fed. Regist.* 49:46324–46331.

Fabro, S., Shull, G., and Brown, N.A. (1982) The relative teratogenic index and teratogenic potency: Proposed components of developmental toxicity risk assessment. *Teratogenesis Carcinog. Mutagen.* 2:61–76.

FDA (Food and Drug Administration) (1970) Food and Drug Administration Advisory Committee on Protocols for Safety Evaluation: Panel on reproduction report on reproduction studies in the safety evaluation of food additives and pesticide residues. *Toxicol. Appl. Pharmacol.* 16:264–296.

FDA (Food and Drug Administration) (1978) Nonclinical laboratory studies. Good laboratory practice regulations. *Fed. Regist.* 43:59986–60020.

Fuyuta, M., Fujimoto, T., and Hirata, S. (1978) Embryotoxic effects of methylmercuric chloride administered to mice and rats during organogenesis. *Teratology* 18:353–366.

Hayes, W.C., Cobel-Geard, S.R., Hanely, T.R., Jr., Murray, J.S., Freshour, N.L., Rao, K.S., and John, J.A. (1981) Teratogenic effects of vitamin A palmitate in Fischer 344 rats. *Drug Chem. Toxicol.* 4:283–295.

Johnson, E.M. (1981) Screening for teratogenic hazards: Nature of the problems. *Annu. Rev. Pharmacol. Toxicol.* 21:417–429.

Johnson, E.M. (1984) A prioritization and biological decision tree for developmental toxicity safety evaluations. *J. Am. Coll. Toxicol.* 3:141–147.

Johnson, E.M. (1987) A tier system for developmental toxicity evaluation based on considerations of exposure and effect relationships. *Teratology* 35:405–427.

Kavlock, R.J., Chernoff, N., and Rogers, E.H. (1985) The effect of acute maternal toxicity on fetal development in the mouse. *Teratogenesis Carcinog. Mutagen.* 5:3–13.

Khera, K.S. (1981) Common fetal aberrations and their teratologic significance: A review. *Fundam. Appl.Toxicol.* 1:13–18.

Khera, K.S. (1984) Maternal toxicity–A possible factor in fetal malformations in mice. *Teratology* 29:411–416.

Khera, K.S. (1985) Maternal toxicity: A possible etiological factor in embryo-fetal deaths and fetal malformations of rabbit-rodent species. *Teratology* 31:129–153.

McLaren, A. and Michie, D. (1960) Congenital runts. In *Ciba Foundation Symposium on Congenital Malformations* (G.E.W. Wolstenholme and C.M. O'Connor, Eds.), pp. 178–198, J. & A. Churchill, London.

Mitchell, S.C., Korones, S.B., and Berendes, H.W. (1971) Congenital heart disease in 56,109 births. Incidence and natural history. *Circulation* 43:323–332.

Mosier, H.D., Jr., Dearden, L.C., Jansons, R.A., Roberts, R.C., and Biggs, C.S. (1982) Disproportionate growth of organs and body weight following glucocorticoid treatment of the rat fetus. *Dev. Pharmacol. Ther.* 4:89–105.

Naeye, R.L. (1980) Effects of maternal nutrition on the outcome of pregnancy. In *Human Embryonic and Fetal Death* (I.H. Porter and E.B. Hook, Eds.), pp. 197–206, Academic Press, Toronto.

NAS (National Academy of Sciences) (1977) *Reproduction and Teratogenicity Tests, Principles and Procedures for Evaluating the Toxicology of Household Substances*, 7, pp. 99–110, National Academy of Sciences, Washington, D.C.

NHW (National Health and Welfare) (1973) *The Testing of Chemicals for Carcinogenicity, Mutagenicity, Teratogenicity*, National Health and Welfare, Canada.

Nishimura, H. (1969) Incidence of malformations in abortions. In *Congenital Malformations* (F.C. Fraser, V.A. McKusick, and R. Robinson, Eds.), pp. 275–283, National Foundation–March of Dimes, Excerpta Medica, New York.

OECD (Organization for Economic Cooperation and Development) (1981) *Teratogenicity*, OECD guidelines for testing of chemicals No. 414, Organization for Economic Cooperation and Development, Paris.

Page, N., Sawbney, D., and Ryon, M.G. (1980) *Proceedings of the Workshop on Subchronic Toxicity Testing*, Oakridge National Laboratory, Interagency Agreement No. 80-D-X0453, U.S. Environmental Protection Agency, Washington, D.C.

Pinto-Machado, J. (1985) External examination of limb positions in near-term mouse fetuses: An experimental study and review of the literature published in teratology. *Teratology* 31:413–423.

Richards, I.D. (1973) Fetal and infant mortality associated with congenital malformations. *Br. J. Prev. Soc. Med.* 27:85–90.

Schardein, J.L. (1983) Teratogenic risk assessment. Past, present and future. *Issues Rev. Teratol.* 1:181–214.

Sterz, H., Sponer, G., Neubert, P., and Hebold, G. (1985) A postulated mechanism of β-sympathomimetic induction of rib and limb anomalies in rat fetuses. *Teratology* 31:401–412.

Task Force of Past Presidents (1982) Animal data in hazard evaluations: Paths and pitfalls. *Fundam. Appl. Toxicol.* 2:101–107.

WHO (World Health Organization) (1984) *Principles for Evaluating Health Risks to Progeny Associated with Exposure to Chemicals during Pregnancy*, Environmental Health Criteria 30, World Health Organization, Geneva.

Wickramaratne, G.A., Killick, M.E., and Appleby, H. (1987) Supernumerary ribs: A transient developmental defect in the rat. In *Pesticide Science and Biotechnology* (R. Greenhalgh and T.R. Roberts, Eds.), pp. 557–559, Blackwell Science Publication, Boston.

Chapter IV

Multigeneration Studies

4.1. Introduction

In the field of toxicology, many disciplines may come into play in making decisions about the inherent properties of a chemical with respect to its potential to cause adverse effects in a particular species of animal. In attempting to predict such adverse effects in humans, the toxicologist is obliged to extrapolate data from experimental animal studies, thus further complicating the process.

In the area of reproduction, the multigeneration study arose as an initial starting point for determining possible adverse effects on the reproductive process. The multigeneration study provides information on the fertility of both male and female parent animals, on the maternal animal during gestation and lactation, on the conceptus in utero, and on the offspring from birth to maturation. There are, however, variations in the conduct of such studies and in extrapolation of study data to the human situation. The following section addresses some of the unresolved and contentious issues in this complex area.

4.2. Test Protocols

Multigeneration and teratogenicity studies are the two major types of investigations that regulatory agencies require to assess the potential of a chemical to induce adverse effects on reproduction or reproductive capacity. In the

United States, the Food and Drug Administration's Division of Pharmacology outlined a procedure to investigate the effect of a chemical fed continuously through three generations (Lehman et al., 1949). The original protocol, somewhat modified, was recommended by the report of the President's Science Advisory Committee (1963) as necessary to establish the safety of pesticides. Since then multigeneration tests have been required by FDA and, subsequently, by EPA as well as other national and international regulatory bodies. These studies include "observations on fertility, size and weight of litter, fetal mortality, teratogenicity, and growth and development of sucklings and weanlings" (Fitzhugh, 1968).

This chapter will not review the history of the changes that have occurred in the original recommendations. Rather, it will examine the scientific basis for the protocols used today and consider some of their strengths and deficiencies. The term "protocols" is used advisedly because the protocol is designed to respond to the question or questions posed. Thus variations in protocols do occur. Multigeneration studies are, in reality, screening studies. Variations between studies depend on the number of variables that were originally proposed for screening in the basic protocol and on the number of other variables to be examined to assist in the resolution of effects that may appear ambiguous in other reproduction studies.

In the protocols currently used, two-generation studies are the normal requirement. A survey of the 70 multigeneration studies evaluated by the FAO/WHO Joint Meeting on Pesticide Residues indicated that in all but two cases the adverse effects observed were at least indicated in the second generation. In the two cases in which effects were first noted in the third generation, the effect was not on reproduction in one case, and in the second it was a minimal lactation-index reduction, which was not further examined. In all other cases, effects were at least indicated during or before the production of the second generation, although in some instances confirmation of the effect required production of a third generation (Clegg, 1979). A similar conclusion was reached in a survey of 34 multigeneration studies (Leeming, N.M., Cozens, D.D., and Palmer, A.K., unpublished data).

Variations in the basic designs of multigeneration reproduction studies are shown in Figure 4-1. It is questionable whether Design 2 and Design 3 differ substantially from each other in terms of the results obtained from using either F_1 litter as the parent animals for the second generation. It has been stated that "the second litters [can] greatly modify the interpretation of results" (Fitzhugh, 1968) or that "studies of effects on fetuses are limited to the second litters, because of variability in reproductive performance in the first mating" (Nisbet and Karch, 1983), but experience has not supported these statements in the majority of cases except for those compounds that accumulate in the test animal or when the first mating has been initiated before full maturity of the parental animals. The data available are insufficient to determine whether the

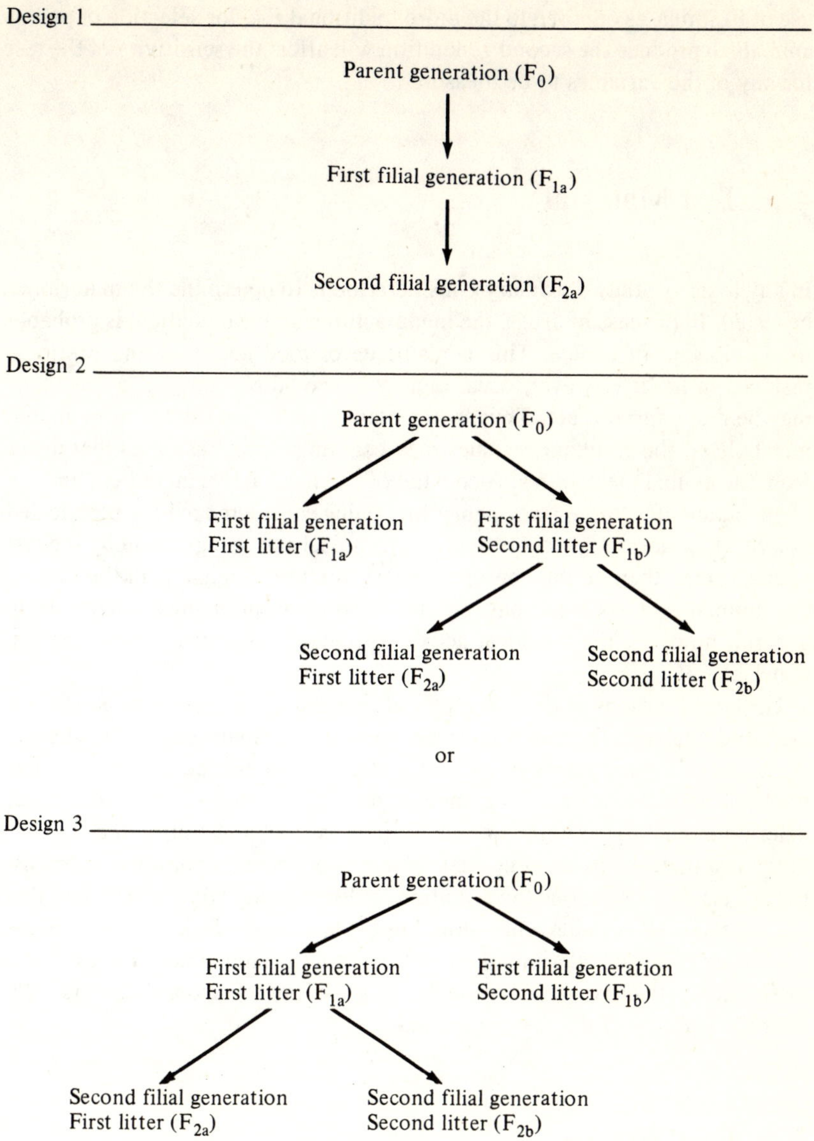

Figure 4-1. Three multigeneration study designs.

use of F_{1a} litter, as opposed to the more traditional F_{1b}, for selection of parent animals to produce the second generation will affect the sensitivity of the test for any of the variables to be measured.

4.3. Test Material

In any toxicity study a primary consideration is to determine the material to be tested. In the case of drugs, the manufactured active ingredient is probably the compound of choice. This is also true of food additives and pesticide residues in food. However, occasionally, in the latter category, a "cocktail" may be more appropriate. Thus, when plant metabolism differs from animal metabolism, the resultant residues may contain plant metabolites that differ from the animal metabolites. A cocktail of technical active ingredient and the plant metabolites (the animal metabolites being considered to have been tested as a result of normal metabolism of the technical active ingredient by the test animals) may then be the appropriate test mixture. Similarly, in the case of contaminants, a cocktail may be the most appropriate test material. In general, however, the technical active ingredient is the usual choice as test material.

The specifications of the test materials elected are an essential part of the available database. The purity and stability of the test material and the identity of the major contaminants in the test material must also be considered. The mode of administration (via gavage, capsule, drinking water, or diet) will determine the information required. With gavage administration the stability of the test material in the proposed solvent is of prime importance. Administration via diet also raises the requirement for stability studies. The fact that the test material is stable when stored in the laboratory does not preclude the possibility of its degradation in or reaction with the animal diet. For example, interaction with components of the diet (e.g., sulfhydryl groups) can radically alter the stability of the test compound.

4.4. Test Species

The species and strain of animal are also basic concerns. In general, the rat is used – mainly because the historical database on various strains of rats is readily available. It is not necessarily the most appropriate species, because rat placentation, relying solely on a yolk-sac placenta during the initial 5 days postimplantation (Beck, 1976), may not necessarily either permit the passage of a same compound as the chorioallantoic placenta of humans or metabolize compounds in a fashion similar to humans. The rat does have the advantage of a short estrus cycle (4–5 days), of being easy to breed, and of having a short gestation period (21–22 days) as well as having a large number of pups per litter. Other rodent species (mainly mice, e.g., Keplinger et al., 1970) have also been utilized, but for these animals there is a relatively restricted historical database on normal reproduction, especially for species having a wide variety of strains and marked variability between these strains (Fraser, 1965).

The cost and duration of the study are also of considerable importance in species selection. Gehring et al. (1973) indicated that the time requirement for a three-generation, two-litter-per-generation reproduction study was comparable for rats and mice (ca 2 years) but longer (30–35 months) if guinea pigs or rabbits were used. The data reported on costs is now outdated, but it can be predicted that the cost of using rabbits or guinea pigs would still exceed the costs involved in using rats or mice. Economics, therefore, favor the use of rats or mice.

Another consideration in selecting the most suitable species for multigeneration studies is the difference encountered in hormonal requirements to maintain pregnancy. In the rodent, pituitary function is essential during the first half of pregnancy and ovarian function for the entire pregnancy. In humans, pituitary function (once conception has occurred) is not required for pregnancy maintenance; ovarian function is only essential during the first trimester. Thus, in rodent studies the occurrence of abortion or resorption may be irrelevant to safety of the test material in humans if the test material interferes with rodent pituitary or ovarian physiology. Under such circumstances, studies in nonhuman primates should be considered.

4.5. Route of Administration

The main purpose of toxicity studies is to determine the safety of compounds for humans. Hence, the route of administration should be chosen to reflect the major route of human exposure. In addition, it should take into account species variations in pharmacokinetics and metabolism, which may be of prime importance because these factors frequently vary according to the route of administration. Thus, for a compound that is metabolized in the liver, oral administration may well generate different results from dermal application (assuming equivalent absorption) because blood levels of parent compound and metabolites reaching the target tissues will differ markedly with the different routes of administration. In general, the optimum administration route is likely to parallel the route of major human exposure. This, however, is a generalization that does not hold true in all cases.

4.6. Method of Administration

Multigeneration studies most often employ oral and occasionally inhalation or dermal routes. In studies using the last two routes, administration is usually designed to follow conditions that reflect occupational exposure patterns. In inhalation studies, an 8-hours-per-day, 5-days-per-week exposure is frequently used. In dermal studies the method and duration of exposure is determined by factors such as potential reexposure (from clothing for instance) and dermal absorption.

The majority of oral studies involve incorporation of the test material in the diet. However, the feeding habits of rodents (little and often) differ markedly from human feeding habits (usually a limited number of feeding times per day). Thus, information on blood levels over a 24-hour period should be reviewed to determine if dietary administration results in blood levels similar to those achieved after bolus administration. If blood-level variations resulting from the two administration methods are excessive, administration by gavage may be a logical alternative. Similarly, if test material stability in the diet is variable or poor, gavage administration may be preferable.

4.7. Dose Levels

Toxicity data available when multigeneration studies are initiated usually relate to adult animals. From these data it is possible to determine those dose levels that, for the adult, are likely to cause either no-observable effect or a minimum toxic effect, neither of which may be relevant to reproduction studies. Because the purpose of multigeneration studies is to investigate effects on reproduction and embryofetotoxicity, it is advisable to perform a pilot study (usually a single-litter study) to select dose levels for the major study that take into account both adult and juvenile toxicity. A pilot study designed to identify adverse effects on libido, fertility, maternal behavior, maternal physiology, and prenatal and postnatal toxicity to the offspring would normally be terminated at weaning. Adverse effects identified in a pilot study may be guiding factors in selection of the highest dose level.

Recently, it has been noted that during the lactation period the food intake of the most widely used species, the rat, increases very markedly (2½ times) during early lactation and decreases to normal intake at about the time of weaning. To maintain a constant intake on a milligram-per-kilogram basis during this period, it has been proposed that the dietary concentration should be reduced appropriately. Although this may have the effect of reducing the maternal toxicity occasionally encountered during lactation, such a reduction increases the problem of extrapolation of data to humans. The human food intake also increases during lactation, and therefore the correction based on rat intake cannot be based on normal intake. Furthermore, the litter size of the rat (even if culling is used) as compared with that of humans would result in a disproportionately lower intake per pup. The possibility, especially for stored compounds, of mobilization of maternal tissue deposits of the test material after a reduction of intake further complicates the problem. Finally, it is probable that the reduced pup exposure would result in the need for supplementary studies on the pups to generate comparative data regarding toxicity to the young animal and time for initiation of activity of enzyme systems. Thus, the problems overcome by reduction of dietary levels to the maternal animals during lactation are likely, at the present state of knowledge, to be exceeded by the problems arising from extrapolation of results to humans.

4.8. Performance of Multigeneration Studies

A number of possible protocols are available for multigeneration studies (FDA, 1982; HWC, 1973; EPA, 1982; OECD, 1983; WHO, 1984). By today's criteria all the protocols should be performed according to good laboratory practice standards. In addition all the protocols indicate that the parent generation (F_0) utilized in producing the first filial generation (F_{1a}) should be pretreated. The duration of this pretreatment period, however, varies from 56–100 days according to the protocol used. The purpose of this pretreatment is to determine effects on libido and fertility of both male and female animals after a suitable period of exposure. It has been proposed that the female should be treated for at least three estrus cycles (usually 15 days) before mating and the male for the duration of one complete spermatic cycle. The duration of the spermatic cycle varies between rat strains (about 48–60 days) (Clermont, 1972; Bowler, 1972). Hence, a 70-day prepairing treatment period may be adequate.

Effects on libido are usually related to levels of circulating testosterone. The levels may be affected early in development or during adulthood, may or may not be cumulative, and may be slow or rapid in occurrence. In fact, to investigate this variable a premating treatment period for F_0 animals cannot be precisely stipulated from our present information. In any event, the considerable variability relating to the duration of the premating treatment period for F_0 parental animals is minimally important in the overall study. The multigeneration study, as far as determining fertility and associated factors involved in initiation of pregnancy are concerned, is at best a screening study. Positive results require additional studies to determine which sex is affected (e.g., mating untreated males with treated females and vice versa, detailed histopathological examination of gonads, and measurement of circulating hormone levels). Furthermore, data are available from the parental animals of the F_2 generation, which have been exposed to the test material from conception and, hence, may be more sensitive to possible adverse effects. On this basis the F_0 generation data on fertility and other variables are in reality back-up data, which are helpful to have but not absolutely essential. Thus, the duration of premating treatment of F_0 parental animals becomes academic in relation to the major purpose of the study.

Pairing of animals may also vary according to the protocol used. Animals may be paired on a 1 female:1 male basis, on a 2 female:1 male basis, etc. The males may be left with the females for one, two, or three estrus cycles. A replacement male may or may not be introduced if there is no evidence of pregnancy after cohabitation with the first male. Onset of pregnancy may be determined by examination of vaginal smears daily or by changes in female body weight. Any combination of these mating procedures is acceptable, but the same combination must be continued throughout all matings in the study unless a sound rationale for change is presented.

In general, a 1:1 pairing regimen is preferred. After the completion of the F_0 breeding, F_1 offspring are used as parental animals to breed the F_2 generation. Cosanguineous matings are avoided to reduce the appearance of possible hereditary defects. This can be difficult even when offspring are produced from 1:1 pairing. It becomes considerably more difficult when two litters have at least one parent in common. For this reason a 1:1 pairing is preferred.

If a pair of animals fails to mate within the pairing period (two or more estrus cycles), it is a common practice to replace the male. In many studies a proven male (one that has already mated) is selected. This, however, biases data on male fertility. If a proven male is used, the rejected male should be paired with an untreated female to assess its fertility. Failure of the male to mate with an untreated female would lead to further investigations (e.g., circulating hormone levels).

Checking of successful pairing (and, hence, the ability to determine the duration of gestation) can provide additional, often useful data. This must, however, be counterbalanced against the possible adverse psychological effects on the maternal animal of daily insertion of a foreign object into the vagina to obtain vaginal smears. Aversion to this tends to be species dependent and is less evident in the rat.

Once pregnancy is established, the only variables normally measured until parturition are body weight and food consumption of the female animal. It is also important to make careful clinical observations of the test animals. At birth, pups are counted and weighed. Postnatal survival is recorded at 1 day and 4 days. At 4 or 5 days postpartum, culling of the litter to a preselected maximum litter size may or may not be performed.

There is controversy regarding the advisability of culling litters. The advocates of culling argue that elimination of competition among pups, reduction of maternal stress, increased uniformity in the pup-sex ratio, and probable increased uniformity between litters are all factors in favor of culling. Conversely, advocates of not culling have suggested that the sensitivity of tests may be increased and that the procedure of culling is unnatural in the test species. Thus, culling may bias the data obtained. Definitive data are rare or lacking to support or refute proposed culling requirements. However, A.K. Palmer (personal communication, 1986) has provided data that support the position of not culling. These data indicate that depending upon the weight or age of the maternal animal at the time of mating, litters from 77% of young dams (less than 250 g) to 91% of older dams (greater than 250-g body weight) would have large enough litters to require culling. (These data also indicate the importance of body weight or age in relation to litter size.) In 2,020 litters at birth, the survival in nonculled litters from day 4 postpartum to weaning (day 21 postpartum) did not show any clear correlation with litter size. The individual pup weight at weaning varied considerably between litters of the same size, whereas mean pup weights showed a slight decrease with increasing litter size of dams nursing more than 6 pups (6 pups, ca 49 g; 8 pups, ca 45.5 g; 10 pups, ca 44.6 g; 12 pups, ca 42 g; 14 pups, ca 40 g). These data indicate that pup losses from litters of 8 pups will bias pup weight much more heavily than similar losses from larger litters. Furthermore, the data indicate that, based on a litter size of 11 pups, approximately one-fourth of the litters did not contain 4 male and 4 female pups, the number normally utilized for culled litters. Thus, the limited available data support the nonculling approach.

In most toxicology studies, once a protocol has been finalized and the study commenced, the protocol should not be subject to major alteration. This is not the case in multigeneration studies. After weaning of the F_{1a} pups, the data should be analyzed as soon as possible to determine the need for a second F_1 litter. An analysis of several variables in more than 50 comparable multigeneration studies indicated that the second litter appeared to be more sensitive than the first litter of the same generation (Clegg and Wandelmaier, 1987). Although the sensitivity of the second litters was greater, there was no case in which effects were observed in the second litters that had not been observed in the first litters. Thus, second litters are not required when the first litters fail to demonstrate any adverse effect on those variables usually evaluated. Neither are second litters required when the observed effect is the induction of nonlethal congenital malformations. Finally, the increased sensitivity of second litters appeared to be limited to postimplantation effects. Thus, if preimplantation effects are the only observed effects, there is no need to produce second litters. Consequently, until the data from the first litter (F_{1a}) are analyzed, the need for a second litter (F_{1b}) cannot be determined. A second litter should be included in the study when the compound is known or likely to be cumulative and intake-output equilibrium has not been attained before the initial mating. In the F_0 generation this will depend upon the length of the premating treatment period and is one reason why, although a pretreatment period of three estrus cycles is the minimum suggested prepairing treatment period, in most multigeneration studies female prepairing treatment periods are commonly identical to those used for males. This further illustrates the need for a reasonably extensive knowledge of the pharmacokinetics and metabolism of the compound before initiating multigeneration studies.

As already indicated, body weight is of major importance with respect to rat reproduction. A.K. Palmer (personal communication, 1986), in an examination of pup weight data from 40 groups of rats comprising 830 litters, has indicated that interlitter variation outweighs intralitter variations by ratio of 4:1. After stating the proviso that experimental design in selecting future parent animals should allow for equal numbers of pups of each sex from each available litter, Palmer continues ". . . in the long run, selection by pre-set criteria would neglect the variation within a group, as well as selection by random methods. If any great store is to be set by the marginal differences between methods, the selection of the lowest weight pup would give the 'widest' variation whilst selection of the high weight pup of each sex would give the 'narrowest' variation which would be the closest to the selection of the 'mean weight pup' (a theoretical choice) in each litter." This probably justifies the current practice of selecting offspring at weaning that are within 3 g of the median weight of the litter after the requirement for selection of equal numbers of pups per sex has been fulfilled, which is rapidly replacing the traditional random pup selection.

At termination of each generation, many protocols require histopathological examination of the parents and of some weanling pups from each dose level. This requirement is questioned because adequate histopathological data are normally available from other studies, at least for the adult animal. It may be necessary, however, to examine the adult reproductive organs, especially in cases of suspected infertility. Similarly, in young animals the organs in which the morphology differs from that in the parent or that are difficult to obtain in older animals (e.g., rat thymus) may warrant examination. Known adult target organs may also warrant investigation to ensure the adequacy of the maximum dose and to determine possible changes in sensitivity in the pregnant female. Should the latter occur, it may be of particular importance in evaluating the study because excessive target organ toxicity may result in abnormal blood levels of the test compound and/or one or more of its metabolites. Overall, therefore, the requirements for histopathological examination of both young and adult tissues must be determined on a case-by-case basis.

4.9. Evaluation of Animal Data

The preceding sections have attempted to indicate the wide spectrum of variations in protocols and in factors affecting the results of multigeneration studies. It is not practical to attempt to address all the possible combinations of variations that would affect the evaluation and extrapolation of data. Hence, this section will briefly outline the basic variables that can be assessed.

Further, as has been indicated, the multigeneration study is basically a screening study for reproductive effects. As a consequence, final evaluation may depend on the results of one or more ancillary studies. The reader is referred to the publication *Methods for Assessing the Effects of Chemicals on Reproductive Functions* (Voulk and Sheehan, 1983) for information on specific studies.

It is axiomatic that to permit a valid evaluation of a multigeneration study all available data must be reported. In many instances this requires that individual data from parental animals on mating schedules, on litters, and on individual pups within each litter are required for each mating in each generation. Further, if variations are introduced either in a normally accepted protocol before initiation of the study or in the selected protocol during the progress of the study, such variations or changes should be fully explained and justified. In the event that the multigeneration study is available only in isolation (i.e., other studies pertinent to evaluation are not available), any relevant information (e.g., chemical characterization of test material and its physical properties, metabolism, pharmacokinetics, and histopathological changes in short or long-term feeding studies) should also be included in the reported data.

Assessment of fertility to a large extent is dependent upon the adequacy of record keeping regarding the details on the identity of each individual male and female paired. The initial 1:1 pairing if unsuccessful requires that future pairings of both the male and female partners be effected. The female should be paired with a proven male. However, within the context of the basic study, the male cannot be mated with a proven female. Thus, an assessment of male fertility frequently requires ancillary studies to confirm or allay suspicions raised by the multigeneration screening study.

The duration of gestation may be affected by the test substances. Thus, if the onset of gestation is determined by vaginal smear techniques, any increase or decrease in gestation duration can be assessed. The normal gestation period in the rat is 21–22 days (with minor variation between strains). Reduction in gestation period also may result in increased fetal incidences of minor variants or malformations, such as delayed ossification, hydronephroses, and hydroureters, because of reduction in the duration of development.

Table 4-1. Historical Control Data on Rat Multigeneration Studies*: Study Design

Study no.	Starting date	Parental generation	Pretreatment period (wk)	Age at mating (wk)	Average female weight at day 0 of gestation	Comment
1	Jan 82	F_0	12	18	286	Not culled
			21	27	321	
		F_{1B}	14	18	274	
			24	28	309	
2	Aug 82	F_0	9	15	267	Not culled
			17	23	306	
		F_{1A}	11	15	274	
			19	23	310	
3	Nov 82	F_0	10	16	291	Not culled
		F_1	12	16	266	
4	Nov 82	F_0	10	16	286	Not culled
		F_1	12	16	272	
5	May 83	F_0	13	19	309	All litters culled to 4 and 4 on day 4
			22	28	333	
		F_{1B}	13	17	305	
			22	26	335	
6	July 83	F_0	10	16	294	Not culled
			19	25	324	
		F_{1B}	10	14	272	
			19	23	312	
7	Aug 83	F_0	9	15	270	Not culled
			18	24	308	
		F_{1A}	11	15	264	
			20	24	307	

*From Huntingdon Research Centre, U.K.; prepared by P. James.

Effects on parturition (e.g., dystocia) are often missed in the normal multigeneration study unless they are predominant. They can be inferred, however, on occasion by a good technician from the condition of the female after parturition. Therefore, immediately after parturition particular attention should be paid to the physical condition of the female, and detailed clinical findings should be included in the report.

Litter size also may indicate a number of possible effects that have occurred during pregnancy. Abortion or partial abortion can be easily missed, especially if untrained animal caretakers are involved or if the female has covertly eaten aborted material. Cannibalization can also occur during or shortly after parturition. Malformed or stillborn pups are the most likely targets for such activity. Again, the observed endpoint may be an apparent reduction in litter

Table 4-2. Historical Control Data on Rat Multigeneration Studies*: Adult Performance–Survival, Fertility, and Care of Young

Study	1		2		3		4		5		6		7	
generation	F_0	F_1	F_0	F_1	F_0	F_1	F_0	F_1	F_0	F_{1B}	F_0	F_{1B}	F_0	F_{1A}
Males														
Group size	32	28	28	24	28	24	28	24	28	24	28	24	28	24
Mortality	0	0	0	0	0	0	0	0	0	0	0	0	0	0
Evidence of mating without issue														
1st mating	1	1	5	3	0	2	2	4	6	4	1	3	1	1
2nd mating	2	1	7	3	–	–	–	–	7	5	1	3	1	0
Either mating	0	0	3	0	–	–	–	–	1	1	0	1	0	0
Females														
Group size	32	28	28	24	28	24	28	24	28	24	28	24	28	24
Mortalities	1	1	0	0	0	0	0	0	1	0	0	0	0	0
Evidence of mating without viable young														
1st mating	1	1	5	3	0	2	2	4	6	4	1	3	1	1
2nd mating	2	1	7	3	–	–	–	–	7	5	1	3	1	0
Either mating	1	1	3	2	–	1[a]	0[a]	2[a]	5	1	1	1	0	0
Delivering with total litter loss														
1st mating	0	0	1	0	1	0	0	1	2	1	0	1	2	1
2nd mating	1	1	0	0	–	–	–	–	1	0	0	0	1	0
Surviving young to weaning														
1st mating	31	27	22	21	27	22	26	19	19	19	27	20	25	22
2nd mating	28	25	21	21	–	–	–	–	19	19	27	21	26	24

*From Huntingdon Research Centre, U.K.; prepared by P. James.
[a] Non-pregnant females remated to males of proven fertility.

size. Other possible causes of reduced litter size may be interference with implantation or resorptions (early or late) due to embryofetotoxicity. Because the female is not available for killing, determination of the exact causes of reduced litter size requires that other studies (e.g., teratogenicity studies or dominant lethal studies) be initiated as deemed necessary.

Increased pup mortality and reduced pup body-weight gain during lactation may result from test exposures of the parent compound and/or its metabolites that may occur during *in utero* development via milk secretions. Ancillary

Table 4-3. Historical Control Data on Rat Multigeneration Studies*: Adult Performance–Time (Days) from Pairing to Mating

Study no.	Generation/ mating	No. rats paired	Number mating successfully at "N" days after pairing: 1	2	3	4	5	6	7	8	9	10	11	12	13	14	15	16	17	18	19	20	Total	Median pairing time (days)
1	$F_{0/1}$	32	7	8	14	1[a]											1						31	3.0
	$F_{0/2}$	32	8	7	11	4																	30	2.5
	$F_{1B/1}$	28	2	11[a]	6	7	1[a]																27	3.0
	$F_{1B/2}$	28	8	5	7	6					1												27	3.0
2	$F_{0/1}$	28	7	7	3	4					1							1					23	2.5
	$F_{0/2}$	28	4	7	3	7																	21	3.5
	$F_{1A/1}$	24	3	5	4	7		1													1		21	3.5
	$F_{1A/2}$	24	7	6	5	2														1			21	2.0
3	$F_{0/1}$	28	9	7[a]	5	4	1				1[a]						1[a]	1[a]					28	2.0
	$F_{1/1}$	24	7	6	2	2		1	1										1[a]	1			22	2.0
4	$F_{0/1}$	28	4	5	7	9														1			26	3.0
	$F_{1/1}$	24	4	5	5	2	1			1			2										20	3.0
5	$F_{0/1}$	27	6	6	3	2	3		1														21	3.0
	$F_{0/2}$	27	7	2	5	4								1[a]		1							20	3.0
	$F_{1B/1}$	24	7	2	3[a]	5[a]													2		1		20	3.0
	$F_{1B/2}$	24	4	4	3	3	2									1		1	1				19	4.0
6	$F_{0/1}$	28	9	4	5	4	2				1		1[a]				1						27	3.0
	$F_{0/2}$	28	6	7	6	7	1																27	3.0
	$F_{1B/1}$	24	5	7	5	3	1																21	2.5
	$F_{1B/2}$	24	3	8	3	5		1													1		21	3.0
7	$F_{0/1}$	28	9	5	4	5[a]				1	1					1						1	27	2.5
	$F_{0/2}$	28	9	2	9	5									1			1					27	3.0
	$F_{1A/1}$	24	9	3	5	6																	23	2.5
	$F_{1A/2}$	24	6	2	10	3					1[a]							1			1		24	3.0

*From Huntingdon Research Centre, U.K.; prepared by P. James.

[a]Indicates absence of indication of mating, but day estimated retrospectively.

Table 4-4. Historical Control Data on Rat Multigeneration Studies*: Sex Ratios

			Total litter size at birth				Live young at day 21				Total dead at weaning	
Study no.	Mode gen.	No. of dams	Males	Females	Total	% Males per litter	Males	Females	Total	% Males per litter	Males	Females
1	$F_{0/1}$	31[1]	6.1	5.6	11.7	52.34	5.9	5.4	11.2	52.16	7	7
	$F_{0/2}$	28	6.7	6.2	12.9	52.33	6.5	6.0	12.5	52.34	6	6
	$F_{1B/1}$	27	5.6	6.6	12.3	45.77	5.3	5.9	11.2	45.71	9	20
	$F_{1B/2}$	25	6.7	5.9	12.6	51.92	6.2	5.4	11.6	52.28	11	12
2	$F_{0/1}$	22	6.2	6.7	12.9	47.9	5.9	6.4	12.3	48.1	6	8
	$F_{0/2}$	21[2]	6.7	6.2	12.9	50.6	6.5	5.9	12.4	52.3	3	7
	$F_{1A/1}$	21	5.3	6.7	12.0	43.2	5.2	6.5	11.7	43.2	2	3
	$F_{1A/2}$	21	5.7	7.0	12.7	44.9	5.4	6.7	12.0	44.9	7	7
3	F_0	27	6.7	6.8	13.6	50.5	6.4	6.0	12.4	52.0	10	22
	F_1	22	5.9	5.5	11.4	51.2	5.5	5.4	10.9	51.4	7	4
4	$F_{0/1}$	26	6.7	6.5	13.2	49.6	6.3	5.9	12.2	51.0	9	15
	$F_{1/1}$	19	6.7	5.5	12.2	55.3	6.4	5.3	11.6	54.7	7	4
6	$F_{0/1}$	27	Data not available									
	$F_{0/2}$	27										
	$F_{1B/1}$	20										
	$F_{1B/2}$	21										
7	$F_{0/1}$	25	6.5	6.4	12.8	49.8	6.0	5.9	11.9	49.7	11	12
	$F_{0/2}$	26	6.8	6.9	13.7	48.8	6.4	6.4	12.8	49.0	10	12
	$F_{1A/1}$	22	6.5	6.7	13.1	49.7	6.0	6.3	12.3	49.5	9	9
	$F_{1A/2}$	24	6.2	7.2	13.3	45.4	6.1	7.0	13.1	45.9	1	4

*From Huntingdon Research Centre, U.K.; prepared by P. James. Includes data only on dams rearing some young to weaning.
Study 5 excluded, culled litters. Superscript numbers indicate number of pups excluded, sex underdetermined at birth.

studies are normally required to determine which of these factors is the major cause. Lactational problems can usually be determined by cross-fostering of pups to untreated females at birth. Although it is possible to determine if the problem lies in the lactation of the treated female with this technique, the question still remains of whether the effects were caused by the chemical constituents in the milk or by reduced lactation in the treated female. This problem is usually addressed by chemical analysis of the milk, if possible. In general, problems arising from lactation of the treated female are suspected when pups of normal birth weight fail to thrive during lactation.

Lack of maternal care can usually be determined by observing the behavior of the mother. It should, however, be recognized that the interpretation of cross-fostering studies may be confused if maternal behavior has not been examined in treated females before undertaking cross-fostering studies.

Table 4-5. Historical Control Data on Rat Multigeneration Studies*: Gestation Periods

Study no.	Gen/ mate	No. of rats	Gestation period 21	22	23	24	Undetermined	Mean duration (days)
1	$F_{0/1}$	31	6	12	12		1	22.2
	$F_{0/2}$	29	1	21	6	1	0	22.2
	$F_{1B/1}$	27	6	11	7	1	2	22.1
	$F_{1B/2}$	26	3	16	6	1	0	22.2
2	$F_{0/1}$	23	11	11	1		0	21.6
	$F_{0/2}$	21	1	18	2		0	22.0
	$F_{1A/1}$	21	7	12	2		0	21.8
	$F_{1A/2}$	21	9	12			0	21.6
3	$F_{0/1}$	28	8	17			3	21.7
	$F_{1/1}$	22	4	12	4		2	22.0
4	$F_{0/1}$	26	5	20	1		0	21.9
	$F_{1/1}$	20	6	12	2		0	21.8
5	$F_{0/1}$	21	4	12	4	1	0	22.1
	$F_{0/2}$	20	3	14	2		1	21.9
	$F_{1B/1}$	20	17	1			2	22.1
	$F_{1B/2}$	19	2	13	3	1	0	22.2
6	$F_{0/1}$	27	9	15	2		1	21.7
	$F_{0/2}$	27	7	19	1		0	21.8
	$F_{1B/1}$	21	3	15	2	1	0	22.0
	$F_{1B/2}$	21	5	13	3		0	21.9
7	$F_{0/1}$	27	7	18	1	0	1	21.8
	$F_{0/2}$	27	1	23	3	0	0	22.1
	$F_{1A/1}$	23	6	15	2	0	0	21.8
	$F_{1A/2}$	24	6	16	1	0	1	21.8

*From Huntingdon Research Centre, U.K.; prepared by P. James.

In utero exposure leading to pup mortality or reduced body-weight gain is often recognizable because of reduced pup weight at birth. It is sometimes advisable to perform studies using untreated males and treated females as parents. More frequently, quantitative and qualitative studies of placental transfer are useful to resolve the basic cause of the observed effects.

In any assessment of biological effects, an essential requirement is a knowledge of the range of variation from normal. Thus, an extensive database that provides information on the mean and the range of values for all variables of interest in the untreated animal is required. In an attempt to establish such a database, historical control data were requested from several laboratories. In many cases, the data could not be provided. It had been anticipated from experience in the examination of many studies that considerable variation would be apparent because of the use of different strains, variations in protocols, variations in environmental conditions in different laboratories, etc.

Table 4-6. Historical Control Data on Rat Multigeneration Studies*: Individual Litter Data

Group			At birth					At 4 days			
Study no.	Gen/ mate	No. of rats	Litter size: Total	Litter size: Viable	% Loss	Litter wt. (g)	Mean pup (g)	Litter size	% Cumulative loss	Litter wt. (g)	Mean pup wt. (g)
					Noncullled litters						
1	$F_{0/1}$	A=B=31	11.7	11.6	0.7	67.4	6.0	11.5	2.4	104.6	9.5
	$F_{0/2}$	A=29	12.5	12.4	0.8	–	–	12.3	1.8	–	–
		B=28	12.9	12.8	0.9	76.1	6.0	12.7	1.8	121.0	9.8
	$F_{1B/1}$	A=B=27	12.3	12.1	0.9	65.3	5.5	11.6	6.0	82.8	7.3
	$F_{1B/2}$	A=26	12.1	11.9	1.7	–	–	11.3	9.7	–	–
		B=25	12.6	12.3	1.7	67.6	5.5	11.8	6.1	93.1	8.0
2	$F_{0/1}$	A=23	12.9	12.7	1.4	–	–	11.9	7.7	–	–
		B=22	12.9	12.7	1.4	69.4	5.5	12.4	3.5	106.1	8.7
	$F_{0/2}$	A=B=21	13.0	12.9	1.1	74.9	5.8	12.5	3.6	117.9	9.4
	$F_{1A/1}$	A=B=21	12.0	12.0	0.0	69.3	5.9	11.8	1.5	108.9	9.4
	$F_{1A/2}$	A=B=21	12.7	12.6	0.8	73.9	5.9	12.4	2.6	116.5	9.5
3	$F_{0/1}$	A=28	13.5	13.4	0.5	–	–	12.6	6.9	–	–
		B=27	13.6	13.5	0.6	75.5	5.7	13.1	3.4	106.1	8.3
	$F_{1/1}$	A=B=22	11.5	11.3	1.6	65.2	5.8	11.0	4.0	102.3	9.4
4	$F_{0/1}$	A=B=26	13.2	13.1	1.0	76.5	5.9	12.5	5.2	108.7	8.8
	$F_{1/1}$	A=20	12.3	12.1	1.1	–	–	11.2	8.1	–	–
		B=19	12.2	12.1	1.1	67.2	5.7	11.8	3.3	104.7	9.0
6	$F_{0/1}$	A=B=27	13.5	13.4	0.5	76.0	5.7	13.1	2.9	112.7	8.8
	$F_{0/2}$	A=B=27	14.0	13.6	2.8	78.0	5.8	13.5	3.3	115.8	8.6
	$F_{1B/1}$	A=21	12.6	12.2	6.9	–	–	12.0	9.1	–	–
	$F_{1B/2}$	B=20	13.0	12.9	2.3	71.6	5.6	12.6	4.5	103.9	8.4
		A=B=21	12.4	12.1	1.7	68.4	5.7	11.8	4.2	100.4	8.7
7	$F_{0/1}$	A=27	12.6	12.0	3.1	–	–	11.4	7.0	–	–
		B=25	12.8	12.6	1.4	69.8	5.6	12.3	3.6	102.5	8.6
	$F_{0/2}$	A=27	13.6	13.0	5.2	–	–	12.7	6.7	–	–
		B=26	13.7	13.5	1.5	73.1	5.9	13.2	3.2	114.4	8.8
	$F_{1A/1}$	A=23	13.3	13.0	2.5	–	–	11.9	9.4	–	–
		B=22	13.1	12.8	2.6	69.3	5.5	12.4	5.3	101.8	8.4
	$F_{1A/2}$	A=B=24	13.3	13.3	0.3	75.2	5.7	13.2	1.2	116.4	8.9
					Culled litters						
								Preculling			
5	$F_{0/1}$	A=21	12.8	12.4	6.8	–	–	11.1	15.4	–	–
		B=19	13.2	13.1	0.8	73.5	5.7	11.8	9.3	100.4	8.6
	$F_{0/2}$	A=20	14.3	13.6	4.1	–	–	12.8	9.5	–	–
		B=19	14.2	14.2	0.0	80.6	5.8	13.5	4.8	109.4	8.2
	$F_{1B/1}$	A=20	11.5	11.5	0.0	–	–	11.2	2.0	–	–
		B=-19	12.1	12.1	0.0	71.4	5.9	11.8	2.1	108.8	9.3
	$F_{1B/2}$	A=B=19	12.5	12.3	1.6	70.7	5.9	11.7	4.7	111.2	9.9

*From Huntingdon Research Centre, U.K.; prepared by P. James.

Table 4-6. (Continued).

At 8 days				At 12 days				At 21 days			
Litter size	% Cumulative loss	Litter wt. (g)	Mean pup wt. (g)	Litter size	% Cumulative loss	Litter wt (g)	Mean pup wt. (g)	Litter size	% Cumulative loss	Litter wt. (g)	Mean pup wt. (g)
Noncculled litters											
11.4	3.1	170.3	15.6	11.3	4.1	235.8	21.7	11.2	4.4	434.4	40.2
12.1	3.1	–	–	12.1	6.3	–	–	12.1	6.5	–	–
12.5	3.2	191.2	15.7	12.5	3.2	272.7	22.4	12.5	8.2	491.3	40.3
11.4	7.1	125.2	11.3	11.3	8.1	187.5	17.1	11.2	8.6	355.0	32.8
11.2	10.7	–	–	11.2	11.0	–	–	11.2	11.0	–	–
11.7	7.2	149.0	13.0	11.6	7.4	232.8	20.4	11.6	7.4	452.3	39.6
11.8	8.0	–	–	11.8	8.0	–	–	11.7	8.6	–	–
12.4	3.9	175.8	14.5	12.4	3.9	262.8	21.6	12.3	4.5	492.3	40.8
12.5	3.9	194.7	15.7	12.5	3.9	292.9	23.6	12.4	4.3	541.0	43.7
11.7	1.9	180.7	15.8	11.7	1.9	262.0	22.9	11.7	1.9	500.2	43.5
12.2	4.1	188.6	15.6	12.0	5.2	279.8	23.4	12.0	5.2	529.9	44.4
12.4	8.0	–	–	12.1	10.4	–	–	11.9	11.3	–	–
12.9	4.5	165.7	13.2	12.5	7.1	223.1	18.2	12.4	8.0	412.3	34.1
10.9	5.2	162.5	15.0	10.9	5.2	231.2	21.3	12.9	5.2	429.1	40.2
12.4	6.3	176.4	14.5	12.2	7.3	252.4	21.1	12.2	7.3	464.6	38.8
11.2	8.1	–	–	11.1	8.9	–	–	11.1	9.3	–	–
11.8	3.3	170.6	14.7	11.7	4.1	241.4	21.1	11.6	4.5	452.7	39.8
12.8	4.6	181.0	14.3	12.8	4.6	265.5	20.9	12.7	5.4	507.7	40.4
13.4	4.2	190.3	14.4	13.3	4.9	283.6	21.5	13.3	4.9	535.3	40.7
11.9	9.7	–	–	11.9	9.7	–	–	11.9	9.7	–	–
12.5	5.1	173.4	14.1	12.5	5.1	268.0	21.8	12.5	5.1	530.1	43.3
11.8	5.8	171.0	15.0	11.8	5.8	267.8	23.5	11.7	6.2	530.4	46.5
11.1	9.4	–	–	11.0	9.6	–	–	11.0	9.6	–	–
12.0	6.2	169.0	14.6	11.9	6.4	253.2	22.0	11.9	6.4	470.1	40.9
12.5	8.7	–	–	12.4	9.2	–	–	12.4	9.5	–	–
13.0	5.2	184.6	14.5	12.9	5.7	274.1	21.6	12.8	6.0	516.7	41.0
11.8	9.7	–	–	11.8	9.7	–	–	11.8	9.9	–	–
12.4	5.6	166.3	13.8	12.4	5.6	247.3	20.5	12.3	5.9	467.0	38.9
13.1	1.5	182.6	14.1	13.1	1.5	274.3	21.1	13.1	1.5	529.4	41.0
Culled litters											
7.1	2.2	–	–	6.7	8.3	–	–	6.7	8.3	–	–
7.5	2.4	117.4	15.6	7.4	3.9	181.5	24.4	7.4	3.9	351.0	47.4
7.5	1.9	–	–	7.4	3.1	–	–	7.4	3.1	–	–
7.8	2.0	122.5	15.6	7.7	3.3	194.2	25.1	7.7	3.3	378.5	48.9
7.6	2.0	–	–	7.5	1.3	–	–	7.5	1.3	–	–
8.0	0.0	135.5	16.9	7.9	1.3	200.6	25.4	7.9	1.3	375.3	50.1
7.6	2.0	131.1	17.4	7.5	2.6	201.7	26.9	7.5	3.3	389.8	52.3

This highlights the inappropriateness of combining data from different laboratories to obtain a basic standard set of historical control data even for one strain. The data provided by P. James of Huntingdon Research Centre (Tables 4-1 through 4-6) illustrate the variations that occur among studies performed within the same laboratory. This information also emphasizes the need for caution and the degree of information that must be available to permit use of any historical control data in interpreting and evaluating multigeneration studies. It illustrates the delicate balance that exists in a single laboratory and how it is subject to change. For this reason an evaluator must be cautious about attributing minor changes to the test compound. Conversely, it indicates the careful judgment the evaluator must use to distinguish between chance variation and variation from toxic effects of the test chemical.

References

Beck, F. (1976) Comparative placental morphology and function. *Environ. Health Perspect.* 18:5–12.

Bowler, K. (1972) The effect of repeated applications of heat on spermatogenesis in the rat: A histological study. *J. Reprod. Fertil.* 28:325–333.

Clegg, D.J. (1979) Animal reproduction and carcinogenicity studies in relation to human safety evaluation. In *Toxicology and Occupational Medicine* (W.B. Deichmann, Ed.), pp. 45–59, Elsevier, North Holland.

Clegg, D.J. and Wandelmaier, F. (1987) An investigation into the desirability of inclusion of a second litter per generation in rat multigeneration studies. In *Pesticide Science and Biotechnology* (R. Greenhalgh and T.R. Roberts, Eds.), pp. 549–552, Blackwell Scientific Publications, Boston.

Clermont, Y. (1972) Kinetics of spermatogenesis in mammals: Seminiferous epithelium cycle and spermatagonial renewal. *Physiol. Rev.* 52:196–236.

EPA (Environmental Protection Agency) (1982) Pesticide Assessment Guidelines, Sub-division F, Hazard Evaluation: Humans and domestic animals, pp. 130–137, Washington, D.C.

FDA (Food and Drug Administration) (1982) *Toxicological Principles for Safety Assessment of Direct Food Additives and Color Additives and Color Additives Used in Food*, No. PB83-170696, pp. 80–107, National Technical Information Service, Washington, D.C.

Fitzhugh, O.G. (1968) Reproduction tests. In *Modern Trends in Toxicology* (E. Boyland and R. Goulding, Eds.), pp. 75–85, Appleton-Century-Crofts, New York.

Fraser, F.C. (1965) Some genetic aspects of teratology. In *Teratology—Principles and Techniques* (J.G. Wilson and J. Warkany, Eds.), University of Chicago Press, Chicago.

Gehring, P.J., Rowe, V.K., and McCollister, S.B. (1973) Toxicology: Cost/time. *Food Cosmet. Toxicol.* 11:1097–1110.

HWC (Health and Welfare Canada) (1973) *The testing of chemicals for carcinogenicity, mutagenicity and teratogenicity*, pp. 137–185, Health and Welfare Canada, Ottawa.

Keplinger, M., Deichman W.B., and Sala, F. (1970) *Pesticide Symposia* (W.B. Deichman, J.L. Radomski, and R.A. Penalver Eds.), p. 125, Halos and Associates, Miami, Florida.

Lehman, A.J., Lang, E.P., Woodard, G., Draize, J.H., Fitzhugh, O.G., and Nelson, A.A. (1949) Procedures for the appraisal of the toxicity of chemicals in food. *Food Drug Cosmet. Law Q.* 4:412.

Nisbet, I.C.T. and Karch, N.J. (1983) *Chemical Hazards to Human Reproduction*, p. 78, Noyes Data Corporation, Park Ridge, New Jersey.

OECD (Organization for Economic Cooperation and Development) (1983) *OECD Guidelines for Testing of Chemicals.* Section 4, *Health Effects*, Guideline 416, May 1983, Organization for Economic Cooperation and Development, Paris, France.

President's Science Advisory Committee (1963) *Use of Pesticides*, U.S. Government Printing Office, Washington, D.C.

Vouk, V.B. and Sheehan, P.J., Eds. (1983) *Methods for Assessing the Effects of Chemicals on Reproductive Functions*, John Wiley & Sons, New York.

WHO (World Health Organization) (1984) *Principles for Evaluating Health Risks to Progeny Associated with Exposure to Chemicals during Pregnancy*, International Programme on Chemical Safety, Environmental Criteria Document, No. 30, World Health Organization, Geneva.

Chapter V

Behavioral-Functional Teratology

5.1. Introduction

During the past decade, there has been intense interest in the development of methods to detect subtle changes resulting from *in utero* and/or perinatal exposure. Because this interest is concerned with protecting society from potential harm, it reflects an advancement in safety assessment (Christian, 1983a; Riley and Vorhees, 1986). Published reports that maternal exposure to lead (Palmisano et al., 1969) and methyl mercury (Matsumoto et al., 1965; Harada, 1968) could produce central nervous system (CNS) disorders in children spurred research activity and regulatory action in behavioral teratology, with an emphasis on neurotoxic effects. Animal models demonstrating neurotoxic effects somewhat similar to those of humans were soon reported (Spyker et al., 1972; Butcher et al., 1973; Spyker, 1975; Hutchings, 1983).

The term *behavioral teratology* was first used when Werboff et al. (1961) and Werboff and Gottlieb (1963) published the results of maternal treatment with meprobamate on the "behavior or functional adaptation of the offspring to its environment." This term, which was initially restricted to CNS alteration, has been extended to include adverse effects on functional development of prenatal origin (Christian and Johnson, 1978, 1979a, 1979b) to include postnatal changes in development and physiological changes in other organ systems. Although still not usually evaluated in standard testing procedures, functional changes were recommended as a part of the U.S. Food and Drug Administration's (FDA) reproductive testing procedures published in 1966. The assessment of functional changes was regarded by Wilson (1973) as one of the four endpoints of developmental toxicity in addition to embryo-fetal death, fetal malformation, and fetal body-weight reduction.

Despite the perception that new techniques are being used, some systems and endpoints have, in fact, been studied since the initial use of developmental toxicology in risk assessment. The major changes resulting in the perception of behavioral-functional teratology being different from classical developmental toxicology have been the emphasis, rediscovery, and/or validation of previously described or suspected changes and the development of methods to detect changes not previously identifiable. For example, the relatively recent development of automated micromethodology has allowed multiple variables to be evaluated using biological fluids from neonates (Christian, 1979), and computerization has allowed automation of procedures to detect sensory threshold responses (Ison, 1978; Ison and Kellog, 1984).

Although the early reproduction guidelines (FDA, 1966) identified the existence of functional alteration, it was not until 1975 that functional alteration was highlighted as a unique, new, subtle endpoint of developmental toxicity. At that time, regulatory agencies of Great Britain, Japan, and Italy began to require behavioral testing as part of their requirements. These countries expanded the original criteria for evaluating reproductive capacity by including functional evaluation in their protocols in addition to the interrelated variables associated with postnatal development (litter viability at birth, growth, and reproductive performance), such as tests for auditory, visual, and behavioral development. The latter tests were required for regulatory purposes as part of Segment 1 studies in the United Kingdom (Committee on the Safety of Medicines, 1974) and Italy and as part of Segments 2 and 3 studies in Japan (MHW, 1975, 1984).

An appraisal of behavioral testing by the U.S. Department of Health, Education and Welfare (Kimmel, 1976; Butcher, 1976; Rodier, 1976; Grant, 1976) and the U.S. Environmental Protection Agency (1984) concluded that because of marked variability in methodology these testing procedures are not sufficiently predictive of human risk to be included in submissions to their agencies. However, studies of these types have been required when the agent is proposed for use in a juvenile population (Danilovitz et al., 1975), when further study might help interpret perinatal findings, or when concern exists about adult neurotoxicity (C.A. Kimmel, personal communication, 1986). The results of two collaborative studies (Buelke-Sam et al., 1985a, 1985b; Tanimura, 1986a, 1986b) indicated the need for further evaluation of the methods and results of such testing.

5.2. Review Articles

Many excellent reviews of testing procedures and critiques of methodology exist and may be consulted for details (Barlow and Sullivan, 1975; Sullivan and Barlow, 1976; Neubert et al., 1977; Rodier, 1978; Adams and Buelke-Sam, 1981; Christian, 1983b; Hutchings, 1983; Jensh, 1983; Lochry et al., 1986; Buelke-Sam and Kimmel, 1979). These reviews point out the need for simple, comprehensive tests that are sensitive to slight alterations in function; that are short, reliable, reproducible, economical, and usable within the existing guidelines for testing; and, above all, that are valid for their ability to predict effects in humans.

5.3. Methods Currently in Use or Under Development

5.3.1. CNS Effects

There is a need for a single, apical behavioral test that can identify multiple effects on integrative functions (Butcher, 1976; Elsner, 1986) but no test exists. However, a battery of tests is available to evaluate different levels of preweaning reflex and cognitive responses, with the emphasis usually placed on redundant, labor-intensive evaluations of developmental landmarks. These endpoints have been demonstrated to be less sensitive than the easily determined variable of pup body weight (Lochry et al., 1984a, 1984b, 1986; Nolen, 1985; Norton, 1986).

When Great Britain, Japan, and Italy enacted regulatory requirements for behavioral evaluation, many pharmaceutical and chemical testing laboratories attempted to develop appropriate testing procedures. These test procedures were usually developed by individuals who were not experienced in behavioral testing. Because the subject was novel, many tests that were not necessarily appropriate were used. A classic summary of testing procedures in use was published in 1979 by Buelke-Sam and Kimmel. This summary included multiple tests used by neurotoxicologists and experimental psychologists in an attempt to comply with regulatory requirements. Many of these procedures continue to be used even though they generally are neither the ideal apical testing paradigm nor appropriate for testing rapidly developing young animals to produce interpretable endpoints (Chester et al., 1985; Lochry et al., 1986).

In general, testing procedures to fulfill regulatory requirements include tests for the assessment of (1) developmental landmarks (physical and reflex development); (2) sensory function and processing; (3) sensorimotor coordination, activity, and exploratory behavior; and (4) learning and retention.

According to Lochry et al. (1984a, 1984b, 1986), F. Sullivan (personal communication, 1984), Lochry (1985), Buelke-Sam et al. (1985a, 1985b) and Nolen (1985), it has been shown that developmental landmarks generally correlate with body weights and that the delayed development evident from these landmarks also is apparent from reduction in body weight of pups. Sexual maturation, evidenced by testicular descent or vaginal patency (hormonally mediated markers), appears to be the only physical landmark that cannot be consistently correlated with body weights (Christian, 1979; Nolen, 1985). This is not to say that pups with relatively low body weights always achieve developmental landmarks more slowly than pups of expected body weights. Rather, the finding demonstrates on a practical basis that pups with low body weights may or may not have additional signs of compromised development. With the advent of highly sensitive electronic balances, it is easier and more cost effective to use body weight as a measure of retarded development, rather than the more variable and less sensitive physical landmarks, when the objective is to identify the no-effect level for a test agent. If the objective is to identify additional (expected) evidence of retarded development, physical and reflex testing may be productive.

Most sensory tests are designed to evaluate responses to visual, auditory, olfactory, or somatosensory stimuli. Tests that are currently being conducted range from the simple snap of a mouse trap for assessing startle response (Jensh, 1983) to sophisticated computer-integrated equipment designed for evaluating responses to multiple stimuli (Ison, 1978; Ison and Kellog, 1984). Although commonly disregarded, the relative sensitivity of a test species to a specific response intended to be assessed should be considered as well as its normal behavior patterns. For example, J. Adams (personal communication, 1977) suggested sensory testing of rats should be conducted under red-light conditions because the species is nocturnal. Although considerable effort has been made to test visual acuity in rats by cliff avoidance (Alder and Zbinden, 1977), the predominant olfactory sense is usually not tested.

Activity tests are purported to be useful in identifying hyperactive animals. Again, the level of sophistication in testing varies widely. Even with electronic testing equipment, including computer integration and data analyses, it remains difficult to predict effects in humans from findings in animals. Hyperactivity resulting from prenatal exposure to alcohol and morphine has been demonstrated in rats (Hutchings, 1983). However, alcohol and morphine were first identified as causing hyperactivity in children, not in laboratory animals. The diagnosis of hyperactivity is based on inappropriate repetitious responses and shortened attention span in children and on increased motor activity in animals. Presumably, testing of other psychopharmacologic agents in similar paradigms could produce data predictive of hyperactivity if the tests were correctly conducted. The greatest problems with these tests include deficient methodology and difficulty in identifying biologically significant observations.

More complex behavioral tests, such as evaluations of exploratory behavior (activity tests), learning, and retention, are difficult to evaluate as part of other ongoing reproductive toxicology studies. The tests are best performed independent of other safety evaluations because they are costly, time consuming, and require invasive procedures consisting of shock and feed deprivation or augmentation in addition to prolonged training procedures.

A multilaboratory study conducted by the National Center for Toxicological Research (NCTR) was reviewed at a poststudy conference at which the results of the collaborative investigation were presented (Buelke-Sam et al., 1985a). The main focus of this study was to investigate the sensitivity of current behavioral tests and their replicability (Adams et al., 1985). The majority of the participants in one of the work groups voted for including behavioral testing as part of existing drug-testing procedures (Kimmel et al., 1985). However, it was also apparent at this conference that no behavioral test had been sufficiently validated or identified as most appropriate for general use (Vorhees, 1985).

5.3.2. Functional Testing Procedures

There is at present an emphasis on CNS-mediated behavioral changes because of active interests in neurotoxic agents, including ethyl alcohol, psychoactive pharmaceuticals, heavy metals, solvents, and petrochemicals. The effects that have been identified correlate highly with the extent and sensitivity of examination. However, the studies should not be limited to the CNS alone.

Many studies for identifying postnatal morphologic and functional alteration have shown that the developing organism is continually at risk and that not only morphologic but biochemical and/or physiological changes can also result from *in utero* and perinatal exposures to chemicals. Some behavioral researchers consider that the available screening tests are appropriate for inclusion in ongoing studies to reveal subtle changes (West, 1985). However, researchers studying postnatal functions generally consider their methods to be a means of identifying behavioral alterations occurring in association with other classic signs of developmental toxicity. Although studies have uncovered functional alterations not shown by classical techniques, to date no chemical has been shown to produce an initial behavioral effect independent of signs of developmental toxicity.

Postnatal organ weight data have been used to demonstrate an increased heart-to-body-weight ratio (Grabowski, 1976). Organ weight data have also produced misleading information (Kavlock et al., 1982). For example, the fetal brain-to-body-weight ratio decreases during late gestation, and overall growth retardation of the fetus results in an increase in the relative brain weight, rather than the expected decrease. Monitoring fetal cardiac function has revealed altered heart rate and electrocardiograms (Grabowski and Turnstall, 1977). Mating of offspring exposed in utero has produced partial or total sterility (Mukherjee et al., 1975). Histological evaluation of rat pups has revealed that prenatal insult can result in morphologic changes, such as delayed maturation of the lung (Armenti et al., 1977; Johnson and Armenti, 1978; Newman et al., 1978), liver, spleen, heart, testis, ovary (Johnson, 1964), and adrenals (Eguchi et al., 1976). Hematological and biochemical analyses have revealed altered hemopoiesis (Amortegui et al., 1976; Hoffman and Campbell, 1977) and derangement of carbohydrate metabolism (DeMeyer et al., 1972; Snell, 1977; Snell et al., 1977), the liver enzyme system (Lucier, 1976), hematocrits, and plasma electrolytes (Chernoff and Grabowski, 1971).

Thus, it should be no surprise that compromised neonates (surviving but with retarded growth) are most likely to have malformations and functional deficits. It has been reported that small babies and babies that appear undernourished at birth are those most likely to have multiple minor malformations and behavioral abnormalities if they survive (Hill, 1986; Hill and Tennyson, 1986).

Mechanistic studies, at present, do not contribute to quantitative risk assessment because the knowledge is fragmentary. It appears that subtle effects are the result of poor study design with evaluation of only the target organ or the behavior or function of interest (Sparber, 1986a, 1986b). If, however, target-organ studies are continued over a period of time, they may provide information useful for risk assessment.

5.3.2.1. Postnatal Alteration of Gastrointestinal, Reproductive, Renal, Hematologic, and Hepatic Function. Rat studies have shown that insult with teratogenic agents during the fetal period can produce physiologic alteration of gastrointestinal, hematologic, renal, and hepatic function (Christian, 1979, 1983b). This work provided a model for evaluation of fetal insult and evaluation of organ systems other than the CNS.

The studies also demonstrated that techniques used in adult pharmacologic and toxicologic testing procedures, if adapted for neonatal testing, could reveal postnatal changes not obvious in the classic teratologic evaluation. Methods were developed for evaluating gastrointestinal transit time, gastric emptying, and rectal emptying in rat offspring exposed *in utero* to maternally toxic dosages of hydroxyurea, actinomycin D, methotrexate, or vitamin A. The gastrointestinal evaluations indicated the importance of evaluating possible sex-related differences in responses. The tests also revealed that the gastrointestinal tract was a target for postnatal change after *in utero* insult.

Simple techniques currently used for evaluating reproductive capacity (examination of smears of vaginal cytology for development and regularity of estrous cycling; mating and reproductive performance; and examination of litter size, growth, and viability of offspring) also are effective for detecting altered reproductive performance of rats exposed *in utero*. Effects reported in teratogen-exposed offspring included delayed vaginal opening, decreased mating performance, abnormal spermatozoa, and decreased fertility.

The advent of microtechnology made it possible to evaluate smaller volumes of body fluids for hematologic and clinical chemistry variables. With this newly developed equipment, Christian (1979) tested neonatal body fluids for changes in hematological, clinical chemistry, and urinalysis values. Blood and urine tests were performed on specimens obtained from the same animals later tested for altered gastrointestinal function. Transient developmental changes in hematology test results (hematocrit and hemoglobin) were observed as were changes in clinical chemistry values (SGOT, SGPT, and BUN) and urinalysis variables. Although these tests were relatively simple to perform, they were either invasive or required killing animals. The microtechnology available at that time permitted testing of relatively small volumes, but collection of blood remains traumatic and invasive. With the advent of improved instrumentation and refined techniques, these tests are now easier to perform, but their usefulness in safety assessment remains to be determined. The work of Kavlock et al. (1982) and Kavlock and Gray (1985) to some extent has provided additional study of some of the biochemical tests with even newer techniques.

5.3.2.2. Alteration of Biochemical Endpoints. In an effort to evaluate fetal toxicity, Kavlock et al. (1982) have studied endpoints of biochemical differentiation of the brain, lungs, liver, and kidneys of late-gestation rat fetuses. They have reported changes associated with fetal growth retardation after exposure to toxic agents. The liver and kidney appear to be more sensitive to insult than do the brain and lungs.

5.3.2.3. Alteration of Renal Functions. Kavlock and Gray (1985) have developed a method to demonstrate a decreased concentrating ability in neonates that have permanent morphological kidney alteration. However, the method has failed to identify effects at doses lower than those causing adverse effects on the morphology of offspring.

5.3.2.4. Alteration of Cardiac Functions. Grabowski and Turnstall (1977) and Grabowski and Payne (1980) developed methods for recording fetal electrocardiograms. The method is relatively easy to perform and requires approximately one-half hour per litter. Although the method requires killing the fetus, it was suggested as an adjunct to classic developmental toxicity testing procedures because the fetus or newborn (Grabowski, 1979) can also be examined for morphologic alteration. This method, although tested in only a small number of animals, is sensitive at very low dose levels.

5.3.2.5. Alteration in Pulmonary Functions. Morphological and biochemical changes in fetal lung have been reported to result from in utero insult by nutritional deprivation (Faridy, 1975), cadmium exposure (Daston and Grabowski, 1979; Daston, 1981), and zinc deficiency (Vojnik and Hurley, 1977). Lungs of rabbit pups exposed to phenobarbital 7–10 days before delivery were shown to have few lamellar bodies and higher opening pressures than controls (Karotkin et al., 1976).

Early neonatal exposure to toxins has also been reported to alter functional lung capacity. Stanley et al. (1975) demonstrated mechanical changes in the lungs of weanling rats exposed to semicarbazide (a lathrogen). The lungs ruptured at lower pressures and, when fixed at 30 cm H_20, were observed to have dilated terminal air sacs and ruptured alveolar walls. Biochemically, there was a decrease in crosslinkage of lung collagen.

Methods to assess postnatal functional alteration after in utero insult have also been developed. Newman et al. (1978) reported that in utero exposure to vitamin A could produce alterations in fetal morphology and physiology, and Newman and Johnson (1983) proposed a model system for the study of respiratory distress in newborns.

5.3.2.6. Reproductive Functional Alteration. Evaluating reproductive capacity is such an integral part of multigeneration testing procedures that it is frequently not recognized as an excellent apical evaluation of behavior, function, and possible anatomical alteration. Standard testing procedures used to screen for pharmacologic toxic effects on reproduction (Christian, 1984; Christian et al., 1986) are appropriate for evaluating postnatal reproductive capacity. In addition to the adverse effects on reproductive physiology during fetal development (Christian, 1979) and previously described for animals with multiple physiological changes, other researchers have reported effects when evaluating rats for specific changes. For example, Bakke et al. (1976) have reported neonatal hypothyroidism as a possible cause of impaired brain development, with concomitant delays in puberty, first estrus, and prolonged estrous cycles. Maternal stress during gestation has been associated with altered sexual behavior in adult male rats (Masterpasqua et al., 1976). Mukherjee et al. (1975) reported reduced fertility or sterility associated with altered RNA synthesis in fetal mouse ovaries after damage with acetyl salicylate or actinomycin D.

5.3.2.7. Immunocompetence Alteration. Changes in immune competence have been reported by Schmidt and Abbott (1983) after in utero exposure of rats to corn oil. Schmidt (1984) has reviewed postnatal effects on immune development and described multiple assays to identify changes following prenatal insult. Assessing defects in the developing immune system has also been discussed (Roberts and Chapman, 1981).

5.4. Current Status of Testing Procedures and Use

At the 1985 International Conference on Neurotoxicology of Selected Chemicals (Tilson and Cranmer, 1986), a close relationship between clinical growth retardation and other functional alteration was emphasized. Hill (1986) reported that children who were small at birth (<5th percentile) or clinically undernourished were more likely to have low IQs, hyperactivity, and learning and cognitive difficulties as well as morphologic changes associated with maternal drug use. Robertson et al., (1986a, 1986b) found increased abortion and, in surviving offspring, malformation, small body weights, and intellectual impairment after antiepileptic treatment during pregnancy. Hill and Tennyson (1986) noted that clinical malnutrition appeared to be a more reliable variable than strict adherence to weight percentiles. In addition, the compromised children were those expected to have behavioral or functional deficits.

Similar observations have been made in animals. Littermates that showed developmental toxicity (death, malformation, or retarded development) also had functional deficits. Many investigators use deficient methodology and continue to report neurologic deficits occurring only at dosages that produce fetal wastage, weight reduction, and other classic signs of develop-mental toxicity (Sparber 1986a, 1986b). Behavioral or functional deficits that occur at toxic levels should be considered inappropriate for human risk analysis because of the importance of maternal nutritional factors and apparently minor (one-day) maternal weight changes.

5.4.1. Great Britain

Great Britain's Committee on Safety of Medicines will continue to request behavioral data and, during a five-year period, will evaluate these data to identify the most useful tests. Developmental landmark data, usually a redundant measure of body weight data (Lochry et al., 1984a, 1984b, 1986; Nolen, 1985; Norton, 1986), are most frequently submitted with few reports on evaluations of learning. According to F. Sullivan (personal communication, 1984) more specific requirements will not be made until the results of the currently available information are assessed.

5.4.2. Japan

Tanimura (1986a, 1986b) concluded that further detailed and long-term collaborative studies were required after the analysis of results of a collaborative study of behavioral teratology in rats conducted in 46 participating laboratories. In this study six test agents (chlorpromazine, ethanol, hydroxyurea, methylaxoxymethanol, phenylalanine, and vitamin A) were assessed using two learning tests (the multiple water T maze and the two-way shuttle box). Behavioral effects were found only when other variables such as growth were impaired. The T maze was reliable but of questionable sensitivity and of ambiguous appropriateness; the shuttle box was considered to be of questionable reliability.

5.4.3. United States

In the United States considerable effort has been expended to standardize test batteries to identify appropriate testing procedures. A document that includes a sample behavioral test battery was made available by the Committee on Postnatal Evaluation of Animals Subjected to Insult During Development (Kimmel, 1976, 1977). Buelke-Sam and Kimmel (1979) summarized behavioral methods for general use. As noted by Lochry et al. (1986), many of the researchers who reported using the tests categorized them incorrectly. Some of the tests were assigned to incorrect or inappropriate behavioral categories, others were not meant to evaluate CNS activity (the variable presumably evaluated by behavioral testing), and some were vaguely described so that their uses could not be determined.

As a result of the obvious deficiencies in the methodology, several collaborative studies were conducted to determine interlaboratory and intralaboratory reproducibility of appropriate testing batteries (Kimmel et al., 1983). The results from the six collaborating U.S. laboratory studies and the proceedings from the workshop-conference at which these results were presented were published (Buelke-Sam et al., 1985a, 1985b).

5.4.3.1. Food and Drug Administration. The requirements, interpretation, and relevancy of observations differ among the five centers within the FDA (1984). In general, the offspring are to be assessed from birth through maturity in multigeneration studies. Assessment of reproductive failure, which does not specifically include behavioral or functional effects, is usually a part of a multigeneration study. Each center within the FDA requires an evaluation of behavioral abnormalities of offspring, but the testing procedures are not specified.

In multigeneration studies, which are usually performed to evaluate direct and indirect food additives, reproductive variables are evaluated in the offspring of exposed parents (the offspring also are exposed to the test agent). These tests are requirements of the Center for Food Safety and Applied Nutrition. The tests are also sometimes required as part of studies submitted to the

Center for Veterinary Medicine (the only center to request assessment of information pertaining to the physiological state of the offspring). These tests require observations on growth, vigor, eye appearance, glossiness of fur, and any type of physical or behavioral abnormality at the time the offspring are weighed and for the period from birth until weaning, approximately 21–28 days postpartum. Color additives are sometimes evaluated by the Center for Devices and Radiological Health. The Center for Drugs and Biologics requires evaluation of overt postnatal abnormalities and behavior as part of Segment 1 (fertility and reproductive-performance studies) and Segment 3 (perinatal and postnatal studies) but does not define which specific tests are needed.

In studies performed for the FDA, the National Center for Toxicological Research (NCTR) recommends evaluating postnatal functional effects when a chemical is suspected to be a functional toxicant in the fetus or adult.

5.4.3.2. Environmental Protection Agency. EPA's proposed guidelines (1984) for risk assessment note that the organ functions of offspring could be adversely affected by chemicals; however, there are no standard procedures for testing these organ functions. It states further that "the determination of functional competence often involves highly specialized training and equipment and is not generally practical for routine testing procedures. Therefore, these approaches may have their greatest application in determining the nature of a suspected alteration in terms of its biological significance and dose-response relationship." In a subsequent EPA publication (1985), the agency has given serious consideration to the establishment of guidelines for behavioral teratology testing, but such guidelines cannot be developed until standardized testing procedures are identified. However, a proposed ruling published for comment in the EPA (1986) included a series of developmental neurotoxicity tests as part of the evaluation of three triethylene glycol ethers. This was the first inclusion of behavioral-functional testing as a proposed requirement under Section 4 of the Toxic Substances Control Act.

5.5. Discussion

The term *behavioral teratology* was coined and generally interpreted in the early 1970s as the study of changes resulting in impaired function after in utero insult. Although the term itself is unfortunate (Johnson et al., 1980) because scientists cannot agree on its meaning or interpretation, it persists. The Behavioral Teratology Society was formed, and by 1975 regulatory submissions to Great Britain, Japan, and Italy included a series of evaluations on the behavior of offspring from dams exposed to test agents.

Two schools of thought and research have evolved on the subject of behavioral-functional teratology. One continues to emphasize behavior, defined as CNS-mediated changes, and is an extension of neurotoxicology and experimental psychology. The other is broader and includes functional alterations of all organ systems, including the CNS. Although the broader term functional-developmental toxicology is now used by some regulatory agencies (EPA, 1984; FDA, 1984), separation of the two disciplines continue. Neurotoxicologists more frequently consider their research to be classified under behavioral teratology, whereas developmental-reproductive toxicologists and physiologists (including classical teratologists) and investigators in immunotoxicology consider their work to be part of functional teratology. If the purpose of a behavioral-functional teratology study is to study mechanisms and developmental changes, all organ systems are subjects for investigation. The most sensitive organ system in which physiological or anatomical alteration occurs at the minimum exposure level should be identified as a target. It is understood that target organs may or may not be species specific. According to the FDA (1970), no single test can be expected to fully define or reveal all possible effects of a test agent.

The choice of testing methods is determined by the intended purpose. Basic mechanistic research is necessary to define changes and to develop methods for their detection. Alternatively, risk assessment is performed on the basis of information demonstrating safe levels of exposure to adults and offspring.

References

Adams, J. and Buelke-Sam, J. (1981) Behavioral assessment of the postnatal animal: Testing and methods development. In *Developmental Toxicology* (C.A. Kimmel and J. Buelke-Sam, Eds.), pp. 233–258, Raven Press, New York.

Adams, J., Buelke-Sam, J., Kimmel, C.A., Nelson, C.J., Reiter, L.W., Sobotka, T.J., Tilson, H.A., and Nelson, B.K. (1985) Collaborative Behavioral Teratology Study: Protocol design and testing procedures. *Neurobehav. Toxicol. Teratol.* 7:579–586.

Alder, S. and Zbinden, G. (1977) Methods for evaluation of physical, neuromuscular, and behavioral development of rats in early postnatal life. In *Methods in Prenatal Toxicology* (D. Neubert, H.J. Merker, and T.W. Kwasigroch, Eds.), pp. 175–185, PSG Publishing Co., Littleton, Massachusetts.

Amortegui, A.J., Feinberg, S.S., and Figallo, E.M. (1976) Postnatal effects of chemically induced intrauterine growth retardation on some hematological values in the rat. *Biol. Neonate* 29:216–221.

Armenti, V.T., Masters, E., Cliver, A., and Johnson, E.M. (1977) Effects of teratogenic insult on fetal lung maturation. *Teratology* 15:30A.

Bakke, J.L., Lawrence, N.L., Robinson, S., and Bennett, J. (1976) Lifelong alterations in endocrine function resulting from brief perinatal hypothyroidism in the rat. *J. Lab. Clin. Med.* 88:3–13.

Barlow, S.M. and Sullivan, F.M. (1975) Behavioural teratology. In *Teratology Trends and Applications* (C.L. Berry and D.E. Poswillo, Eds.), pp. 103–120, Springer-Verlag, Berlin.

Buelke-Sam, J. and Kimmel, C., (1979) Development and standardization of screening methods for behavioral teratology. *Teratology* 20:17–29.

Buelke-Sam, J., Kimmel, C.A., and Adams, J., Eds. (1985a) Design considerations in screening for behavioral teratogens: Results of the Collaborative Behavioral Teratology Study. *Neurobehav. Toxicol. Teratol.* 7:537–789.

Buelke-Sam, J., Kimmel, C.A., Adams, J., Nelson, C.J., Vorhees, C.V., Wright, D.C., St. Omer, V., Korol, B.A., Butcher, R.E., Geyer, M.A., Holson, J.F., Kutscher, C.L., and Wagner, J.F. (1985b) Collaborative Behavioral Teratology Study: Results. *Neurobehav. Toxicol. Teratol.* 7:591–624.

Butcher, R.E. (1976) Behavioral testing as a method for assessing risk. *Environ. Health Perspect.* 18:75–78.

Butcher, R.E., Scott, W.J., Kazmaier, K., and Ritter, E.J. (1973) Postnatal effects in rats of prenatal treatment with hydroxyurea. *Teratology* 7:161–165.

Chernoff, N. and Grabowski, C.T. (1971) Responses of the rat fetus to maternal injections of adrenaline and vasopressin. *Br. J. Pharmacol.* 43:270–278.

Chester, A., Hallesy, D., and Andrew, F. (1985) Behavioral methods in reproductive and developmental toxicology. *Neurobehav. Toxicol. Teratol.* 7:745–752.

Christian, M.S. (1979) Postnatal functional teratology resulting from fetal insult, Dissertation, Thomas Jefferson University, Philadelphia, Pennsylvania.

Christian, M.S. (1983a) Statement of problem. In *Assessment of Reproductive and Teratogenic Hazards* (M.S. Christian, W.M. Galbraith, P. Voytek, and M.A. Mehlman, Eds.), pp. 1–4, Princeton Scientific Publishers, Princeton, New Jersey.

Christian, M.S. (1983b) Postnatal alterations of gastrointestinal physiology, hematology, clinical chemistry and other non-CNS parameters. In *Teratogenesis and Reproductive Toxicology* (E.M. Johnson and D.M. Kochhar, Eds.), pp. 263–286, Springer-Verlag, Berlin.

Christian, M.S. (1984) Reproductive toxicity and teratology evaluations of naltrexone. *J. Clin. Psychiatry* 45:7–10.

Christian, M.S., Hoberman, A.M., and Lochry, E.A. (1986) Clinical observations and effects predicted in developmental and reproductive toxicity evaluations of naltrexone. *Neurotoxicology* 7(1):353–354, abstract.

Christian, M.S. and Johnson, E.M. (1978) Prenatal alteration of postnatal gastrointestinal function. *Teratology* 17:44A.

Christian, M.S. and Johnson, E.M. (1979a) Postnatal alteration of non-CNS physiology evaluated in rats treated with teratogens during the fetal period. *Teratology* 19:23A.

Christian, M.S. and Johnson, E.M. (1979b) Functional teratology of the gastrointestinal tract resulting from fetal insult in the rat. 18th Annual Meeting–Society of Toxicology, p. A119, abstract no. 237.

Committee on the Safety of Medicines (1974) *Notes for Guidance on Reproduction Studies*, Committee on the Safety of Medicines, Department of Health and Social Security, London, Great Britain.

Danilovitz, M.S., Lightkep, G.E., and Mosher, A.H. (1975) Pediatric study of chronic morphine treatment in rats. *Toxicol. Appl. Pharmacol.* 33:169–170.

Daston, G.P. (1981) Effects of cadmium on the prenatal ultrastructural maturation of rat alveolar epithelium. *Teratology* 23:75–84.

Daston, G.P. and Grabowski, C.T. (1979) Toxic effects of cadmium on the developing rat lung. I. Altered pulmonary surfactant and the induction of respiratory distress syndrome. *J. Toxicol. Environ. Health* 5:973–983.

DeMeyer, R., Verellen, G., and Gerard, P. (1972) Study of carbohydrate metabolism in the newborn rat as a tool for evaluating effects of drugs administered during pregnancy. In *Drugs and Fetal Development* (M.A. Klingberg, A. Abramavici, and J. Chemke, Eds.), pp. 83–96, Plenum Press, New York.

Eguchi, Y., Arishima, K., Morikawa, Y., and Hashimoto, Y. (1976) Changes in the weight of the adrenal glands and in the concentration of plasma corticosterone in perinatal rats after prenatal treatment with estradiol benzoate. *J. Endocrinol.* 69:427–431.

Elsner, J. (1986) New concepts in assessing behavioral toxicology. In *Safety Evaluation and Regulation of Chemicals*, 3, (F. Homburger, Ed.), pp. 62–69, Karger, Basel.

EPA (Environmental Protection Agency) (1984) Proposed guidelines for the health assessment of suspect developmental toxicants. *Fed. Regist.* 49:46324–46331.

EPA (Environmental Protection Agency) (1985) *Scientific Rationale for the Selection of Toxicity Testing Methods. II. Teratology, Immunotoxicology and Inhalation Toxicology* (M.G. Ryon and D.S. Sawhney, Eds.), pp. 30–36, National Technical Information Service, ORNL-6094; EPA-560/6-84-004, Springfield, Virginia.

EPA (Environmental Protection Agency) (1986) Triethylene glycol monomethyl, monethyl, and monobutyl ethers; proposed test rule. *Fed. Regist.* 51:17883–17892.

Faridy, E. (1975) Effect of maternal malnutrition on surface activity of fetal lungs in rats. *J. Appl. Physiol.* 39:535–540.

FDA (Food and Drug Administration) (1966) *Guidelines for Reproduction Studies for Safety Evaluation of Drugs for Human Use*, U.S. Food and Drug Administration, Washington, D.C.

FDA (Food and Drug Administration) (1970) Advisory Committee on Protocols for Safety Evaluations: Panel on Reproduction Report on Reproduction Studies in the Safety Evaluation of Food Additives and Pesticide Residues. *Toxicol. Appl. Pharmacol.* 16:264–269.

FDA (Food and Drug Administration) (1984) *Final Report of Task Force on Reproductive and Developmental Toxicology*, U.S. Food and Drug Administration, Washington, D.C.

Grabowski, C.T. (1976) Postnatal effects on rats of prenatal exposure to carbon monoxide. *Teratology* 13:23A.

Grabowski, C.T. (1979) An analysis of the causes of perinatal deaths induced by prenatal exposure to Mirex. *Teratology* 19:27A.

Grabowski, C.T. and Payne, D.B. (1980) An electrocardiographic study of cardiovascular problems in Mirex-fed rat fetuses. *Teratology* 22:167–177.

Grabowski, C.T. and Turnstall, A.C. (1977) An electrocardiographic study of rat fetuses treated with trypan blue. *Teratology* 15:32A.

Grant, L.D. (1976) Research strategies for behavioral teratology studies. *Environ. Health Perspect.* 18:85–94.

Harada, Y. (1968) Infantile Minamata disease. In *Minamata Disease*, pp. 73–91, Univ. Japan Study Group of Minamata Disease.

Hill, R. (1986) Commonly prescribed drugs and fetal development, malnutrition and neurological development in the fetus and child. *Neurotoxicology* 7(1):358–359, abstract.

Hill, R.M. and Tennyson, L.M. (1986) Maternal drug therapy: Effect on fetal neonatal growth and neurobehavior. *Neurotoxicology* 7(2):121–139.

Hoffman, D.J. and Campbell, K.I. (1977) Postnatal toxicity of carbon monoxide after pre- and postnatal exposure. *Toxicol. Lett.* 1:147–150.

Hutchings, D.E. (1983) Behavioral teratology: A new frontier in neurobehavioral research. In *Teratogenesis and Reproductive Toxicology* (E.M. Johnson and D.M. Kochhar, Eds.), pp. 207–235, Springer-Verlag, Berlin.

Ison, J. (1978) Reflex inhibition and reflex elicitation by acoustic stimuli differing in abruptness of onset and peak intensity. *Anim. Learn. Behav.* 6:106–110.

Ison, J. and Kellog, C.K. (1984) Behavioral analysis of sensory function. *Teratology* 19:36–37A.

Jensh, R.P. (1983) Behavioral testing procedures: A review. In *Teratogenesis and Reproductive Toxicology* (E.M. Johnson and D.M. Kochar, Eds.), pp. 171–206, Springer-Verlag, Berlin.

Johnson, E.M. (1964) A histologic study of postnatal vitamin B12 deficiency in the rat. *Am. J. Pathol.* 44:73–83.

Johnson, E.M. and Armenti, V.T. (1978) Postnatal effects of prenatal insult on lung development in the rat. *Anat. Rec.* 190:432–433.

Johnson, E.M., Staples, R.E., and Hoar, R.M. (1980) Behavioral teratogens—Unfortunate terminology. *Neurobehav. Toxicol. Teratol.* 2:iv, commentary.

Karotkin, E.H., Kido, M., Redding, R.A., Cashore, W.J., Douglas, W., Stern, L., and Oh, W. (1976) The inhibition of pulmonary maturation in the fetal rabbit by maternal treatment with phenobarbital. *Am. J. Obstet. Gynecol.* 124:529–531.

Kavlock, R.J. and Gray, J.A. (1985) Functional evaluation of the kidney in reproductive toxicology screens. J. *Am. Coll. Toxicol.* 4(6), abstract no. 168.

Kavlock, R.J., Chernoff, N., Rogers, E., Whitehouse, D., Carver, B., Gray, J., and Robinson, K. (1982) An analysis of fetotoxicity using biochemical endpoints of organ differentiation. *Teratology* 26:183–194.

Kimmel, C.A. (1976) Behavioral teratology: Overview. *Environ. Health Perspect.* 18:73.

Kimmel, C.A., Ed. (1977) *Final Report of the Committee on Postnatal Evaluation of Animals Subjected to Insult during Development.* National Institute of Environmental Health Sciences, Research Triangle Park, North Carolina.

Kimmel, C.A., Buelke-Sam, J., and Adams, J. (1985) Collaborative Behavioral Teratology Study: Implications, current applications and future directions. *Neurobehav. Toxicol. Teratol.* 7:669–673.

Kimmel, C.A., Buelke-Sam, J., Adams, J., Reiter, L.W., Sobotka, T.J., and Tilson, J.A. (1983) An interlaboratory comparison of selected methods in a behavioral teratology protocol. *Teratology* 27:57–58A.

Lochry, E.A. (1985) Application of behavioral evaluations in reproductive screens: Methodological and practical considerations. *J. Am. Coll. Toxicol.* 4(6), abstract no. 167.

Lochry, E.A., Hoberman, A.M., and Christian, M.S. (1984a) Positive correlation of pup body weight with other commonly used developmental landmarks. *Teratology* 29:44A.

Lochry, E.A., Hoberman, A.M., and Christian, M.S. (1984b) Comparative sensitivity of commonly used developmental landmarks and pup body weight. Presented at the European Teratology Society Meeting, Koningshof, The Netherlands, September 1984.

Lochry, E.A., Hoberman, A.M., and Christian, M.S. (1986) Standardization and application of behavioral teratology screens. In *Safety Evaluation and Regulation of Chemicals*, 3, (F. Homburger, Ed.), pp. 49–61, S. Karger, Basel.

Lucier, G.W. (1976) Perinatal development of conjugative enzyme systems. *Environ. Health Perspect.* 18:25–34.

Masterpasqua, F., Chapman, R.H., and Lore, R.K. (1976) The effects of prenatal psychological stress on the sexual behavior and reactivity of male rats. *Dev. Psychobiol.* 9:403–411.

Matsumoto, H.G., Koya, G., and Takeuchi, T. (1965) Fetal Minamata disease. A neuropathological study of two cases of intrauterine intoxication by a methyl-mercury compound. *J. Neuropathol. Exp. Neurol.* 24:563–574.

MHW (Ministry of Health and Welfare) (1984) *Information on the Guidelines of Toxicity Studies Required for Applications for Approval to Manufacture (Import) Drugs . . . (Part I)*, Notification No. 118, Japanese Pharmaceutical Affairs Bureau, Ministry of Health and Welfare, Tokyo, Japan.

MHW (Ministry of Health and Welfare) (1975) *On Animal Experimental Methods for Testing the Effects of Drugs on Reproduction*, Pharmaceutical and Supply Bureau, Ministry of Health and Welfare, Tokyo, Japan.

Mukherjee, A.B., Chan, M., Waite, R., Metzger, M.I., and Jaffee, S.J. (1975) Inhibition of RNA synthesis by acetyl salicylate and actinomycin-D during early development in the mouse. *Pediatr. Res.* 9:652–657.

Neubert, D., Merker, H.-J., and Kwasigroch, T.E., Eds. (1977) Evaluation of postnatal manifestation of prenatally-induced lesions. In *Methods in Prenatal Toxicology: Evaluation of Embryotoxic Effects in Experimental Animals, Teratology Workshop*, pp. 166–200, PSG Publishing Co., Littleton, Massachusetts.

Newman, L.M. and Johnson, E.M. (1983) Abnormal lung function induced by prenatal insult. In *Teratogenesis and Reproductive Toxicology* (E.M. Johnson and D.M. Kochhar, Eds.), pp. 237–261, Springer-Verlag, Berlin.

Newman, L.M., Johnson, E.M., and Armenti, V.T. (1978) Effect of prenatal vitamin A on neonatal respiratory frequency and 0_2 consumption in the Long-Evans' rat. *Teratology* 17:44A.

Nolen, G.A. (1985) Developmental landmarks—Useful or not? *J. Am. Coll. Toxicol.* 4(6), abstract no. 162.

Norton, S. (1986) Behavioral changes in preweaning and adult rats exposed prenatally to low ionizing radiation. *Toxicol. Appl. Pharmacol.* 83:240–249.

Palmisano, P.A., Sneed, R.C., and Cassady, G. (1969) Untaxed whiskey and fetal lead exposure. *J. Pediatr.* 75:869–872.

Riley, E.P. and Vorhees, C.V. (1986) *Handbook of Behavioral Teratology*, pp. 1–522, Plenum Press, New York.

Roberts, D.W. and Chapman, J.R. (1981) Concepts essential to the assessment of toxicity to the developing immune system. In *Developmental Toxicology* (C.A. Kimmel and J. Buelke-Sam, Eds.), pp. 167–189, Raven Press, New York.

Robertson, L.D., Swaiman, K.F., and Ptacek, L.J. (1986a) Methodology for follow up of children at risk for growth deficiencies due to antiepileptic drug (AED) exposure during pregnancy: A population based study. *Neurotoxicology* 7(1):359, abstract.

Robertson, L.D., Swaiman, K.F., and Ptacek, L.J. (1986b) Fetal anticonvulsant drug exposure: A population based study. *Neurotoxicology* 7(2):413–419.

Rodier, P.M. (1976) Critical periods for behavior anomalies in mice. *Environ. Health Perspect.* 18:79–83.

Rodier, P.M. (1978) Behavioral teratology. In *Handbook of Teratology*, Vol. 4, *Research Procedures and Data Analysis* (J.G. Wilson and F.C. Fraser, eds.), pp. 397–428, Plenum Press, New York.

Schmidt, R.R. (1984) Altered development of immunocompetence following prenatal or combined prenatal-postnatal insult: A timely review. *J. Am. Coll. Toxicol.* 3:57-72.

Schmidt, R.R. and Abbott, P.K. (1983) Altered postnatal mitogenic responsiveness of adult rat splenic lymphocytes following *in utero* exposure to corn oil. *Teratology* 27:411–416.

Snell, J. (1977) Effect of prenatal exposure to methylmercury on perinatal carbohydrate metabolism in the rat, p. 13, International Congress of Toxicology, Toronto, Canada.

Snell, K., Ashby, S.L., and Barton, S.J. (1977) Disturbances of perinatal carbohydrate metabolism in rats exposed to methylmercury *in utero*. *Toxicology* 8:277–283.

Sparber, S.B. (1986a) Developmental effects of narcotics. *Neurotoxicology* 7(1):357, abstract.

Sparber, S.B. (1986b) Developmental effects of narcotics. *Neurotoxicology* 7(2):325–348.

Spyker, J.M. (1975) Behavioral teratology and toxicology. In *Behavioral Toxicology* (B. Weiss and V.G. Laties, Eds.), pp. 311–349, Plenum Press, New York.

Spyker, J.M., Sparber, S.B., and Goldberg, A.M. (1972) Subtle consequences of methylmercury exposure: Behavioral deviations in offspring of treated mothers. *Science* 177:621–623.

Stanley, N.N., Alper, R., Cunningham, E.L., Cherniack, N.S., and Kefalides, N.A. (1975) Effects of a molecular change in collagen on lung structure and mechanical function. *J. Clin. Invest.* 55:1195–1201.

Sullivan, F.M. and Barlow, S.M. (1976) Behavioural teratology: Some problems in assessment of post-natal development and behaviour. In *Advances in the Detection of Congenital Malformation* (E.B. van Julsingha, J.M. Tesh, and G.M. Fara, Eds.), pp. 87–93, European Teratology Society, Kent, England.

Tanimura, T. (1986a) Collaborative studies on behavioral teratology by 46 laboratories in Japan. *Neurotoxicology* 7(1):370, abstract.

Tanimura, T. (1986b) Collaborative studies on behavioral teratology in Japan. *Neurotoxicology* 7(2):35–46.

Tilson, H.A. and Cranmer, J.M., Eds. (1986) Neurotoxicology in the fetus and child. *Neurotoxicology* 7(2).

Vojnik, C. and Hurley, L.S. (1977) Abnormal prenatal lung development resulting from maternal zinc deficiency in rats. *J. Nutr.* 107:862–872.

Vorhees, C.V. (1985) Comparison of the Collaborative Behavioral Teratology Study and Cincinnati behavioral teratology test batteries. *Neurobehav. Toxicol. Teratol.* 7:625–633.

Werboff, J. and Gottlieb, J.S. (1963) Drugs in pregnancy: Behavioral teratology. *Obstet. Gynecol. Surv.* 18:420.

Werboff, J., Gottlieb, J.S., Dembicki, E.L., and Havlena, J. (1961) Postnatal effect of antidepressant drugs administered during gestation. *Exp. Neurol.* 3:542–555.

West, G.L. (1985) Considerations of sensory motor coordination tests during growth and development. *J. Am. Coll. Toxicol.* 4(6), abstract no. 163.

Wilson, J.G. (1973) *Environment and Birth Defects*, p. 28, Academic Press, New York.

Chapter VI

In Vitro Developmental Toxicity Assays

6.1. Introduction

The development and applications of in vitro systems for safety evaluation and environmental risk studies on chemicals (i.e., toxicity testing) have been, in large part, limited to the field of genetic toxicology. Recent advances in molecular and cellular differentiation have provided new insights into the mechanisms of embryonic development and new opportunities to analyze the causes and effects of teratogenic insult. For example, observations of the role of glucocorticoids (Pratt et al., 1984a, 1984b; Pratt, 1985), epidermal growth factor (Grove and Pratt, 1984), NAD (Midura et al., 1985), the composition of the extracellular matrix (Swalla and Solursh, 1984; Solursh et al., 1984), and the possible influence of secretory products of embryonic ectoderm cells on mesenchymal cell differentiation (Solursh, 1984b) have contributed to greater understanding of the factors controlling skeletal system morphogenesis. Investigations such as those using the nematode *Caenorhabditis elegans* (Sulston and Horvitz, 1977) have developed and extended the concept of cell lineages in the process of postembryonic growth. Mechanistic and morphologically based investigations of the types just described proceed apace. Their results suggest that mechanistic analyses, rather than traditional morphological studies of laboratory animals, may meet future prerequisites of testing for teratogenic potential.

This prediction is supported by the recent development of a variety of teratogen screening systems (Homburger and Goldberg, 1985; Purchase and

Conning, 1986). These assays include those using dissociated mammalian cells cultured in vitro; mammalian embryos or embryonic organs, avian embryonic cells, or organs in vitro; and other submammalian forms such as *Xenopus* spp., *Drosophila*, and *Hydra*. These new approaches to analyzing the potential embryotoxic and teratogenic hazards of drugs and environmental agents are the subject of this chapter.

6.2. Assay Systems Using Submammalian Species

Several assay systems using invertebrates or nonmammalian vertebrates have been proposed for teratogen screening. These systems exhibit potential advantages over those using mammalian embryos and tissues or the cells derived from them. These advantages include reductions in laboratory mammal usage, complexity of assay systems, and cost as well as ready availability of test species. At the same time possible disadvantages should also be noted. Among these is the issue of phylogenetic disparity in developmental mechanisms. It could be argued that because teratogen screening assays are aimed at questions of human health and safety, the assays should employ models that reflect, as closely as possible, human developmental patterns. At present the extent to which the effects of teratogens on invertebrate and lower vertebrate development resemble those observed in mammalian development is largely unknown. Nonetheless, basic research, such as that using nematodes (Sulston and Horvitz, 1977), sea urchins (Karp and Solursh, 1985), *Hydra attenuata* (Johnson et al., 1984), and *Drosophila* (Buzin and Bournias-Vardiabasis, 1984), continues to contribute significantly to an understanding of the mechanisms of differentiation.

6.2.1. Invertebrate Models

Basic research observations, such as those just alluded to, have led to the development of a number of teratogen screening assays. Principal among the invertebrate assay systems currently available are the *Hydra* (Johnson et al., 1982) and *Drosophila* embryo cell (Bournias-Vardiabasis and Teplitz, 1982) assay systems. The *Hydra* system described by Johnson et al. (1982) uses both fully formed (adult) organisms and dissociated cells as target systems. Treatments are administered in the aqueous milieu. Toxicity to the adult is monitored by morphological evaluation of the adult hydroid, and embryonic toxicity is judged by the effects of treatment on reaggregation of hydroid-derived cells. The concentrations of the test substance resulting in patent adult toxicity and inhibition of cellular reaggregation are compared to obtain a ratio that can be used to construct adult vs. embryo toxicity rankings. A limited number of correlation studies have been conducted to validate these systems.

Another system that uses invertebrate organisms to evaluate teratogenic potential has been described by Bournias-Vardiabasis and Teplitz (1982). In primary cultures of *Drosophila* embryo cells, these investigators determined the effects of test treatments on cell differentiation and tissue formation in vitro, using an automated image-analysis system. These effects included those related to organogenesis, such as cell death, cell migration and recognition, and cell fusion. This system can be used to screen sera from treated laboratory animals (Bournias-Vardiabasis and Teplitz, 1982), and proteratogen activation can be provided by rat or *Drosophila* microsomes (Bournias-Vardiabasis and Flores, 1983). A limited amount of validation data has been gathered for this assay system.

6.2.2. Nonmammalian Vertebrate Models

Two lower-vertebrate-based assay systems for teratogen screening are currently available. One of these uses intact early-stage frog embryos as the test target, and the other employs organ or cell culture of chick embryo limb buds.

The frog embryo assay system (Dumont et al., 1983) uses *Xenopus laevis* embryos to measure effects of treatments with presumed teratogens. The test chemicals are dissolved in the aquatic environment of the developing embryos, and the effects are usually evaluated after 96 hours. An advantage of this system is that embryonic death can be readily distinguished from more subtle effects on development. Thus, both lethal concentrations (LC_{50}) and developmental effect concentrations (EC_{50}) can be quantitatively determined for each test material. These data can be used to estimate relative teratogenic risk, with large LC_{50}/EC_{50} ratios implying higher teratogenic risk than ratios approaching unity (e.g., Schultz et al., 1985).

Another nonmammalian system or family of systems employs chick embryos *in ovo*, chick embryo limb buds in organ culture, or chick embryo limb bud cells in vitro (see Caplan, 1972; Royal et al., 1980; Solursh, 1984a, 1984b; and Hattori and Ide, 1984, for some recent investigations using chick-embryo-derived models). With the exception of *in ovo* investigations, most of the basic methodologies for using chick embryos in teratogenic studies are similar to those using rodent embryos. Because of these methodological similarities and the basic similarities in the in vitro observations obtained using chick and rodent tissue organ and cell cultures, mammalian systems are emphasized in the following sections.

6.3. Assay Systems Using Mammalian Species

The development of in vitro assays for teratogenicity using intact embryos, organs, or cells derived from mammalian species has advanced rapidly. Several different endpoints, ranging from morphological alterations in cultured rodent embryos to biochemical or radiochemical analyses of cultured rodent limb-bud cells, have been proposed for evaluation. Many of these recently developed assay systems attempt to bridge the gap between morphological observations of terata induced in utero and the mechanisms of their induction as analyzed in vitro. Test systems using mammalian-derived target materials have the potential advantage of reducing the impact of teleological arguments concerning the use of submammalian species in teratogen screening. Most mammalian assay systems presently available use rodent-derived materials; therefore, they are also of questionable relevance to human beings.

Teleological arguments aside, these current in vitro assays use cultured cell lines, cell strains, or cultured embryos so that laboratory animals are not used at all, or if they are, their numbers are greatly reduced compared with in vivo assays. Although this may be a favorable result for alternative test development, the use of mammalian tissues does not resolve all the questions relating to the significance of in vivo vs. in vitro observations that exist along with the phylogenetic questions raised earlier. For example, issues such as the role of maternal metabolism and transplacental transport have not yet been fully addressed in the course of in vitro assay development.

6.3.1. Embryo Culture Models

Several laboratories have been working to develop reliable methods for rodent whole-embryo culture, and much recent progress has been made in basic and applied research (see Kochhar, 1975, for an historical perspective). Both rat and mouse embryos have been cultured successfully in vitro and have been used in recent teratogenicity research (Knudsen et al., 1985; Pratt et al., 1984a, 1984b; Ellington and New, 1980; Steele et al., 1983a, 1983b). Most embryo culture assays for teratogenic potential use endpoints closely related to those evaluated in in vivo studies. Because of this, the results of embryo culture assays can be directly compared with in vivo observations, thus providing a unique opportunity for validation. Assay development continues (Sadler and New, 1981; Steele et al., 1983a, 1983b; Flint et al., 1984; Satish et al., 1985) and the use of embryo culture techniques in screening is uncertain pending further validation.

6.3.2. Embryo Limb-Bud Organ and Cell Culture Models

The use of rodent limb buds in organ culture has resulted in numerous observations regarding the mechanisms of teratogenesis (see Kochhar, 1975). In vitro assays using limb buds generally use morphological endpoints; thus, as is the case for whole-embryo culture, the results can be compared with those of in vivo assays. Thus, cultured limb buds have been used to investigate defective gene expression (Kochhar, 1985), and they have potential applications to a variety of questions, including those relating to cartilagenous differentiation in limb-bud cell cultures.

From an applied perspective of teratogen screening assay development, dissociated limb buds in cell culture have received recent attention. This is largely because of the types of basic research to which cultured limb-bud cells assays are amenable. In this regard, cultured limb-bud cells have been used to investigate the role of retinoid receptors (Kwarta et al., 1985), of the cytoskeleton (Zanetti and Solursh, 1984), and of cell donor-strain specificity (Campbell et al., 1985) in teratogenesis.

Teratogen assays using limb-bud cells have been described that measure proteoglycan accumulation using a specific stain (alcian blue stain, Wilk et al., 1980; Flint et al., 1984) or radiosulfate incorporation (Guntakatta et al., 1984). Measurements of proteoglycan synthesis by the latter technique are accompanied by an estimate of the treatments' effects on the viability and growth of the target cells, thus allowing separation of "teratogenic" and cytotoxic effects. The database for validating cultured limb-bud cells for teratogen assays is growing rapidly.

6.3.3. Embryonic Palatal Cell Models

Embryonic palatal cells, like embryonic limb-bud cells, undergo a cartilagenous phase of differentiation in vitro. Both mouse and human palatal mesenchymal cell populations have been used in mechanistic studies (Kim et al., 1984; Yoneda and Pratt, 1981), and a database has been reported for the use of a human palatal cell strain in teratogen screening (Pratt and Willis, 1985). The assay system measures inhibitions of colony formation with or without a microsomal enzyme activation system (Pratt and Willis, 1985). This assay has the advantages that it is not technically difficult, it uses an established cell line, and it generates reproducible results. However, the assay does not measure an element of differentiation per se; therefore, its validation ultimately rests on correlation studies.

6.3.4. Cultured-Cell Attachment Models

Another assay system that uses mammalian cells has been described by Braun and his coworkers (1979). This assay measures the effects of teratogen treatments on the ability of cultured mouse ovarian tumor cells to attach to lectin (concanavalin A)-coated plastic surfaces. A considerable amount of correlation data has been collected for this assay (Braun et al., 1982; Braun and Horowicz, 1983; Braun and Weinreb, 1984), and a system for activating thalidomide has been reported (Braun and Weinreb, 1985). A potential advantage of this assay system is its use of an established cell line. Thus, experimental variance should be reduced, resulting in good reproducibility. Also, except for the use of radiothymidine, the assay does not require sophisticated equipment and, therefore, should be readily available to interested investigators. However, like the embryonic palatal cell assay, this system does not directly measure perturbations of differentiation but relies on the putative similarity between lectin attachment and cell-to-cell and cell-to-extracellular-matrix recognition and adhesion.

6.4. Role of In Vitro Assays for Teratogenicity

Current research into the basic biology of embryogenesis will continue to provide the basis for new developments in teratogen testing. Nonetheless, it is apparent that the testing methodologies currently available have applications as screening tools. Some of these tests are being used during product development and may find significant use in the area of product liability. In any case, as is true for genetic toxicology testing, in vitro teratogen assays may be best applied in a test battery scheme. The design and success of such test batteries is, however, largely dependent upon the development of a validation database. Approaches to the validation of in vitro methods in teratogen testing have been suggested (Kimmel et al., 1982; Kimmel, 1985). The key element in validating these in vitro assays systems is their accuracy in predicting human teratogenicity. In this regard the reference compounds of choice are the known human teratogens. However, the major teratogenesis database was derived from rodent studies (see Shepard, 1980); thus, from a statistical point of view, validation by correlation alone would likely require use of the rodent database. Uncertainty resulting from the use of such correlation data could be resolved if the in vitro test endpoint could be shown by process validation to be mechanistically allied with the processes of in situ teratogenic insult.

The question of mechanisms as it relates to in vitro test development is central to the process of validation and to test-battery design because there are likely multiple mechanisms involved in the induction of terata (Wilson, 1978).

For example, assay systems that measure perturbations in chondrogenesis (e.g., limb-bud assays) may yield useful information concerning a test material's potential for inducing skeletal abnormalities but are less likely to identify teratogens that act on embryonic tissues of nonmesenchymal origin.

It should be stressed that major hurdles remain in the development of predictive in vitro developmental toxicity assays. Very little is understood about the biology of normal, much less abnormal, development. At present, the in vitro systems discussed in this chapter provide promise for mechanistic rescarch and possibly for thc sclcction from groups of structurally and functionally related compounds those that possess the least potential for developmental toxicity. At the present state of development, however, these in vitro assays cannot and should not be used for human risk assessment.

References

Bournias-Vardiabasis, N. and Flores, J. (1983) Drug metabolizing enzymes in *Drosophila melanogaster*: Teratogenicity of cyclophosphamide *in vitro*. *Teratogenesis Carcinog. Mutagen.* 3:255–262.

Bournias-Vardiabasis, N. and Teplitz, R.L. (1982) Use of *Drosophila* embryo cell cultures as an *in vitro* teratogen assay. *Teratogenesis Carcinog. Mutagen.* 2:333–341.

Braun, A.G., Buckner, C.A., Emerson, D.J., and Nichinson, B.B. (1982) Quantitative correspondence between the *in vivo* and *in vitro* activity of teratogenic agents. *Proc. Natl. Acad. Sci. U.S.A.* 79:2056–2060.

Braun, A.G., Emerson, D.J., and Nichinson, B.B. (1979) Teratogenic drugs inhibit tumour cell attachment to lectin-coated surfaces. *Nature* 282:507–509.

Braun, A.G. and Horowicz, P.B. (1983) Lectin-mediated attachment assay for teratogens: Results with 32 pesticides. *J. Toxicol. Environ. Health* 11:275–286.

Braun, A.G. and Weinreb, S.L. (1984) Teratogen metabolism: Activation of thalidomide and thalidomide analogues to products that inhibit the attachment of cells to concanavalin A coated plastic surfaces. *Biochem. Pharmacol.* 33:1471–1477.

Braun, A.G. and Weinreb, S.L. (1985) Teratogen metabolism: Spontaneous decay products of thalidomide and thalidomide analogues are not bioactivated by liver microsomes. *Teratogenesis Carcinog. Mutagen.* 5:149–158.

Buzin, C.H. and Bournias-Vardiabasis, N. (1984) Teratogens induce a subset of small heat shock proteins in *Drosophila* primary embryonic cell cultures. *Proc. Natl. Acad. Sci. U.S.A.* 81:4075–4079.

Campbell, M.A., Brown, K.S., Hassell, J.R., Horigan, E.A., and Keeler, R.F. (1985) Inhibition of limb chondrogenesis by a *Veratrum* alkaloid: Temporal specificity *in vivo* and *in vitro*. *Dev. Biol.* 111:464–470.

Caplan, A.I. (1972) Effects of a nicotinamide-sensitive teratogen 6-aminonicotinamide on chick limb cells in culture. *Exp. Cell Res.* 70:185–195.

Dumont, J.N., Schultz, T.W. Buchanan, M., and Kao, G. (1983) Frog embryo teratogenesis assay: *Xenopus* (FETAX)–A short-term assay applicable to complex environmental mixtures. In *Symposium on the Application of Short-Term Bioassays in the Analysis of Complex Environmental Mixtures*, III (M.D. Waters, S.S. Sandhu, J. Lewtas, L. Claxton, and S. Nesnow, Eds.), pp. 393–405, Plenum Press, New York.

Ellington, S.K. and New, D.A. (1980) *In vitro* development of the rat parietal yolk sac. *J. Reprod. Fertil.* 60:377–382.

Flint, O.P., Orton, T.C., and Ferguson, R.A. (1984) Differentation of rat embryo cells in culture: Response following acute maternal exposure to teratogens and non-teratogens. *J. Appl. Toxicol.* 4:109–116.

Grove, R.I. and Pratt, R.M. (1984) Influence of epidermal growth factor and cyclic AMP on growth and differentiation of palatal epithelial cells in culture. *Dev. Biol.* 106:427–437.

Guntakatta, M., Matthews, E.J. and Rundell, J.O. (1984) Development of a mouse embryo limb bud cell culture system for the estimation of chemical teratogenic potential. *Teratogenesis Carcinog. Mutagen.* 4:349–364.

Hattori, T. and Ide, H. (1984) Limb bud chondrogenesis in cell culture, with particular reference to serum concentration in the culture medium. *Exp. Cell Res.* 150:338–346.

Homburger, F. and Golberg, A.M. (1985) *In Vitro Embryotoxicity and Teratogenicity Tests*, S. Karger, Basel.

Johnson, E.M., Gabel, B.E., and Larson, J. (1984) Developmental toxicity and structure/activity correlates of glycols and glycol ethers. *Environ. Health Perspect.* 57:135–139.

Johnson, E.M., Gorman, R.M., Gabel, B.E.G., and George, M.E. (1982) The *Hydra attenuata* system for detection of teratogenic hazards. *Teratogenesis Carcinog. Mutagen.* 2:263–276.

Karp, G.C. and Solursh, M. (1985) *In vitro* fusion and separation of sea urchin primary mesenchyme cells. *Exp. Cell Res.* 158:554–557.

Kim, C.S., Lauder, J.M., Joh, T.H., and Pratt, R.M. (1984) Immunocytochemical localization of glucocorticoid receptors in midgestation murine embryos and human embryonic cultured cells. *J. Histochem. Cytochem.* 32:1234–1237.

Kimmel, G.L. (1985) *In vitro* tests in screening teratogens: Considerations to aid the validation process. *Prog. Clin. Biol. Res.* 163C:259–263.

Kimmel, G.L., Smith, K., Kochhar, D.M., and Pratt, R.M. (1982) Overview of in vitro teratogenicity testing: Aspects of validation and application to screening. *Teratogenesis Carcinog. Mutagen.* 2:221–229.

Knudsen, T.B., Elmer, W.A., and Kochhar, D.M. (1985) Elevated rate of DNA synthesis and its correlation to cAMP- phosphodiesterase activity during induction of polydactyly in mouse embryos heterozygous for *Hemimelia-exta toe* (Hm^X). *Teratology* 31:155–166.

Kochhar, D.M. (1985) Cellular expression of a mutant gene (cmd/cmd) causing limb and other defects in mouse embryos. *Prog. Clin. Biol. Res.* 163C:131–144.

Kochhar, D.M. (1975) The use of *in vitro* procedures in teratology. *Teratology* 11:273–287.

Kwarta, R.F., Jr., Kimmel, C.A., Kimmel, G.L., and Slikker, W., Jr. (1985) Identification of the cellular retinoic acid binding protein (cRABP) within the embryonic mouse (CD-1) limb bud. *Teratology* 32:103–111.

Midura, R.J., Cherney, B.W., and Caplan, A.I. (1985) The relationship of nicotinamide adenine dinucleotide to the chondrogenic differentiation of limb mesenchymal cells. *Dev. Biol.* 111:232–242.

Pratt, R.M. (1985) Receptor-dependent mechanisms of glucocorticoid and dioxin-induced cleft palate. *Environ. Health Perspect.* 61:35–40.

Pratt, R.M., Kim, C.S., and Grove, R.I. (1984a) Role of glucocorticoids and epidermal growth factor in normal and abnormal palatal development. *Curr. Top. Dev. Biol.* 19:81–101.

Pratt, R.M., Perry, E.L., Chapman L.M., and Goulding, E.H. (1984b) Glucocorticoid teratogenesis in mouse whole embryo culture. *Teratology* 30:71–81.

Pratt, R.M. and Willis, W.D. (1985) *In vitro* screening assay for teratogens using growth inhibition of human embryonic cells. *Proc. Natl. Acad. Sci. USA* 82:5791–5794.

Purchase, I. and Conning, D. (1986) International conference on in vitro toxicology. *Food Chem Toxicol.* 24:601–656.

Royal, P.D., Sparks, K.J. and Goetinck, P.F. (1980) Physical and immunochemical characterization of proteoglycans synthesized during chondrogenesis in the chick embryo. *J. Biol. Chem.* 255:9870–9878.

Sadler, T.W. and New, D.A. (1981) Culture of mouse embryos during neurulation. *J. Embryol. Exp. Morphol.* 66:109–116.

Satish, J., Pratt, B.M., and Sanyal, M.K. (1985) Differential dysmorphogenesis induced by microinjection of an alkylating agent into rat conceptuses cultured *in vitro*. *Teratology* 31:61–72.

Schultz, T.W., Dumont, J.N., and Epler, R.G. (1985) The embryotoxic and osteolathyrogenic effects of semicarbazide. *Toxicology* 36:183–198.

Shepard, T.H. (1980) *Catalog of Teratogenic Agents*, 3rd ed., Johns Hopkins University Press, Baltimore.

Solursh, M. (1984a) Ectoderm as a determinant of early tissue pattern in the limb bud. *Cell Differ.* 15:17–24.

Solursh, M. (1984b) The migratory capacity of myogenic cells *in vitro*. *Dev. Biol.* 102:509–513.

Solursh, M., Jensen, K.L., Zanetti, N.C., Linsenmayer, T.F., and Reiter, R.S. (1984) Extracellular matrix mediates epithelial effects on chondrogeneais *in vitro*. *Dev. Biol.* 105:451–457.

Steele, C.E., New, D.A.T., Ashford, A., and Copping, G.P. (1983a) Teratogenic action of hypolipidemic agents: An *in vitro* study with postimplantation rat embryos. *Teratology* 28:229–236.

Steele, C.E., Trasler, D.G., and New, D.A.T. (1983b) An *in vivo/in vitro* evaluation of the teratogenic action of excess vitamin A. *Teratology* 28:209–214.

Sulston, J.E. and Horvitz, H.R. (1977) Post-embryonic cell lineages of the nematode *Caenorhabditis elegans*. *Dev. Biol.* 56:110–156.

Swalla, B.J. and Solursh, M. (1984) Inhibition of limb chondrogenesis by fibronectin. *Differentiation* 26:42–48.

Wilk, A.L., Greenberg, J.H., Horigan, E.A., Pratt, R.M., and Martin, G.R. (1980) Detection of teratogenic compounds using differentiating embryonic cells in culture. *In Vitro* 16:269–276.

Wilson, J. (1978) *Handbook of Teratology*, vol. 4, *Research Procedures and Data Analysis*, Plenum Press, New York.

Yoneda, T. and Pratt, R.M. (1981) Mesenchymal cells from the human embryonic palate are highly responsive to epidermal growth factor. *Science* 213:563–565.

Zanetti, N.C. and Solursh, M. (1984) Induction of chondrogenesis in limb mesenchymal cultures by disruption of the actin cytoskeleton. *J. Cell Biol.* 99:115–123.

Chapter VII

Statistical Methods for Developmental Toxicity Studies

7.1. Introduction

As detailed in the preceding chapters of this monograph, developmental toxicity may be expressed in terms of a number of possible reproductive or teratological abnormalities resulting from exposure to a particular toxicant. Developmental toxicity may be evaluated in studies involving exposure of one or both parents before conception, exposure of the conceptus during pregnancy, or perinatal exposure of the offspring. Multigeneration studies may also be used to assess the cumulative effects of exposure over several generations.

The most commonly used protocol for assessing developmental toxicity in laboratory animals involves the administration of a test substance to pregnant animals (usually mice, rats, or rabbits) during the period of major organogenesis (FDA, 1970; EPA, 1985). Developmental effects are determined by monitoring maternal responses throughout pregnancy and examining the dam as well as the contents of the uterus just before parturition. Other protocols may invoke exposure periods of only one or several days to investigate periods of particular sensitivity for the induction of abnormalities in specific organs.

Experimental designs should generally include a high dose that produces a minimal level of maternal toxicity expressed in terms of marginally reduced body weight or organ-specific toxicity. A low dose that demonstrates no observed adverse effects is also recommended along with at least one intermediate dose level. A concurrent control group treated with the vehicle used for agent administration should also be employed. The test animals should be selected on the basis of species, strain, age, weight, and health status, and they should be randomized into dose groups to avoid bias and to provide a basis for performing valid statistical tests.

Oser and Oser (1956) defined four primary indices of interest in a reproduction study. The *fertility index* is the fraction of attempted matings that result in a successful pregnancy. The *gestation index* is the percentage of pregnant females that deliver viable litters containing at least one live offspring. The *viability index* is the fraction of live offspring that survive for at least four days after birth. The *lactation index* is the fraction of those animals alive at 4 days of age that survive until the time of weaning (about 21 days after birth in rats and mice).

Other related indices may also be of interest. These include average litter size and the average pup weight. These indices differ from the four indices proposed by Oser and Oser (1956) in that they are not defined in terms of simple percentages and thus require different methods of statistical analysis (Gad and Weil, 1986).

This chapter addresses the statistical analysis of binary data indicating the presence or absence of a particular developmental anomaly. With independent binary outcomes, as with the accumulation of the number of successful matings in calculating the fertility index, standard statistical methods of analysis apply. Specifically, fertility rates in exposed females may be compared with those in unexposed controls, tests for increasing trends in infertility with increasing dose may be conducted, and dose-response models in which fertility rates are expressed as a function of dose may be fitted (Gad and Weil, 1986; Gart et al., 1987).

Because developmental toxicity data frequently include observations on animals from the same litter, the binary outcomes cannot always be regarded as independent. This may result in *litter effects*, which represent the tendency for litter mates to respond more similarly than animals from different litters (Mantel, 1969; Weil, 1970). A number of factors may be responsible for this variability in response, including the handling of the animals, genetic factors, and prenatal environment.

In analyzing such data, it is necessary to take into account the within-litter correlation by using the variation between litters rather than variation between animals as the basis for between-group comparisons. Below we discuss in detail the statistical consequences of this in general terms and outline specific methods of analysis that may be used with such data.

In section 7.2, statistical tests for comparing the rate of occurrence of adverse effects in exposed animals relative to unexposed controls are described. Each fetal effect (resorption, fetal death, fetal weight or length reduction, or individual malformation) is analyzed independently. All of these procedures take into account potential intralitter correlation in the responses observed among littermates. These procedures may be used to establish a no-observed-effect level, or NOEL, at which the rate of occurrence of adverse effects in exposed animals is not significantly greater than that in unexposed controls. The determination of a NOEL is important for risk assessment purposes in that it serves as a reference point to which a suitable safety or uncertainty factor may be applied in order to arrive at human safety standards (Hogan and Hoel, 1982).

Statistical models that may be used to describe the dose-response relationship for developmental effects are discussed in section 7.3. These models differ from those often employed for carcinogenic effects in that they provide for litter effects. These dose-response models provide a quantitative description of the manner in which the rate of occurrence of developmental effects varies with dose and provide a means of estimating risk at doses not included in the experimental protocol.

7.2. Comparisons Between Exposed and Unexposed Groups

In the presence of positive intralitter correlation, standard methods of statistical analysis that ignore the litter structure will tend to underestimate the standard error of the difference between the overall response rates in two different treatment groups and, hence, to overstate the statistical significance of any observed differences (Haseman and Hogan, 1975). Gladen (1979) showed that the use of standard chi-squared tests that ignore intralitter correlation can result in inflated false-positive rates. In this regard, Rao and Scott (1981) have shown that the correct asymptotic null distribution of the usual chi-squared statistic for comparing several treatment groups is in fact a weighted sum of independent chi-square random variables with weights related to the intralitter correlation within each group. The use of the Fisher-Irwin exact test for comparing two treatment groups, ignoring the litter structure, also has been shown to result in somewhat inflated false-positive rates in the presence of positive intralitter correlation (Krewski et al., 1984). With negative intralitter correlation, however, the standard tests would be valid in the sense that the actual false-positive rate would tend to be less than the nominal rate.

To avoid these problems, statistical methods that take litter structure into account should be employed. Let n_{ij} denote the size of the jth litter in the ith treatment group ($j = 1, \ldots, m_i$; $i = 1, \ldots, I$), and let x_{ij} denote the number of these animals affected. Conditional on the n_{ij}, Cochran (1943) assumed that the sample proportion $\hat{p}_{ij} = x_{ij}/n_{ij}$ have mean

$$E(\hat{p}_{ij} \mid p_{ij}) = p_{ij}$$

and variance

$$V(\hat{p}_{ij} \mid p_{ij}) = p_{ij}(1 - p_{ij})/n_{ij},$$

where the p_{ij} are considered to be held constant. If the p_{ij} are actually independent random variables with mean

$$E(p_{ij}) = \mu_i$$

and variance

$$V(p_{ij}) = \sigma_i^2 > 0,$$

we have

$$E(\hat{p}_{ij}) = \mu_i$$

and

$$\sigma_{ij}^2 = V(\hat{p}_{ij}) = \frac{\mu_i(1 - \mu_i)}{n_{ij}} + \sigma_i^2 \left(1 - \frac{1}{n_{ij}}\right).$$

Note that the first term in the above formula represents the variation that would occur in the absence of any litter effects, whereas the second term reflects the between-litter variation associated with such effects. Cochran proposed a weighted analysis of variance of the observed proportions $\hat{p}_{ij}$ as a means of assessing treatment difference, with estimates of σ_{ij}^2 used to obtain the weights (see also Kleinman, 1973).

Gladen (1979) used a more general model with $V(\hat{p}_{ij}) = f(n_{ij})$ for some general function f. Although the natural estimator

$$\hat{p}_i = x_{i.}/n_{i.} = \sum_{j=1}^{m_i} x_{ij} \Big/ \sum_{j=1}^{m_i} n_{ij}$$

of μ_i is unbiased, Gladen proposed the unbiased jackknife estimator

$$p_{Ji} = m_i\hat{p}_i - \frac{(m_i - 1)}{m_i} \sum_{j=1}^{m_i} \hat{p}_{i(-j)},$$

where $\hat{p}_{i(-j)} = (x_{i.} - x_{ij})/(n_{i.} - n_{ij})$ denotes the estimator of μ_i omitting the jth litter in group i. The jackknife estimator of the variance of each p_{Ji} is given by

$$\nu_{Ji} = \frac{m_i - 1}{m_i} \sum_{j=1}^{m_i} \hat{p}_{i(-j)} - \frac{1}{m_i} \sum_{j=1}^{m_i} p_{i(-j)}{}^2.$$

To compare two groups with m_1 and m_2 litters, respectively, the statistic

$$t_j = \frac{\hat{p}_{J1} - \hat{p}_{J2}}{(\nu_{J1} + \nu_{J2})}^{1/2}$$

will then approximate a t distribution with $m_1 + m_2 - 2$ degrees of freedom under the null hypothesis H_0: $\mu_1 = \mu_2$, provided m_1 and m_2 are sufficiently large.

Frangos and Stone (1984) and Southward and Van Ryzin (1972) investigated, among other approaches, the estimator

$$\bar{p}_i = \sum_{j=1}^{m_i} \hat{p}_{ij}/m_i,$$

which is the average of the litter-specific proportions in group i. The variance of p_i can be estimated by

$$v_i = \sum_{j=1}^{m_i} (p_{ij} - p_i)^2/\{m_i(m_i - 1)\}.$$

For comparison of two groups with m_1 and m_2 litters, this would lead to a standardized test statistic

$$z = (p_1 - p_2)/(v_1 + v_2)^{1/2},$$

which asymptotically follows a standard normal distribution. The small sample behavior of z has not been investigated. For a single group, however, Frangos and Stone (1984) demonstrated in a small simulation study that confidence intervals based on the $\bar{p}_i$ and on a modification of an estimator proposed by Southward and Van Ryzin (1972) outperform the jackknife confidence intervals in certain circumstances.

Another nonparametric technique has been considered by Haseman and Soares (1976). In particular, they showed that comparing the litter-specific proportions in two treatment groups using the nonparametric Wilcoxon test (Hollander and Wolfe, 1973, p. 68) is in many circumstances a sufficiently accurate and powerful statistical procedure.

Schreiner and Steelman (1977) assess the significance of overall effects by a Kruskal–Wallis test and then employ Dunn's multiple comparison procedure for pairwise testing. These nonparametric techniques can be modified to take ties into account (Haseman and Hoel, 1974; Stucky and Vollmar, 1976; Lin and Haseman, 1976).

A more parametric approach to modeling litter effects was used by Williams (1975), who assumed that p_{ij} follows a beta distribution with parameters α_i, $\beta_i > 0$. In this case, $\mu_i = \alpha_i/(\alpha_i + \beta_i)$ and $\sigma_i^2 = \mu_i(1 - \mu_i)\, \pi_i/(\alpha + \pi_i)$, where $\pi_i = (\alpha_i + \beta_i + 1)^{-1} > 0$ provides a measure of the degree of association between littermates. A recent extension to the beta-binomial model given by Prentice (1986) allows for the possibility of negative intralitter correlation. Williams proposed the use of beta-binomial likelihood ratio methods to test the null hypothesis H_0: $\mu_1 = \mu_2$; $\pi_1 = \pi_2$ against a general alternative. Williams also considered H_0: $\mu_1 = \mu_2 = \mu$ with $\pi_1 = \pi_2 = \pi$ fixed. Because

$\pi \to 0$ as $\mu \to 0$ or 1, however, the assumption of constant intralitter correlation may not be appropriate when the response rates μ_1 and μ_2 differ appreciably. Based on a simulation study, Shirley and Hickling (1981) concluded that for typically encountered litter sizes, the nonparametric Wilcoxon test was preferable to the beta-binomial likelihood ratio test (see also Haseman and Kupper, 1979).

A different approach has been suggested by Kupper and Haseman (1978). They assume that fetuses in the same litter tend to have an inherent relationship to one another. Thus, the assumption of mutual independence of the outcomes within a litter, which usually leads to a binomial within-litter model, is altered by applying a correction factor that depends on the covariance between two Bernoulli trials within one litter. Kupper and Haseman (1978) demonstrate that this correlated binomial model is, in a sense, an extension of the beta-binomial model as discussed by Williams (1975) in that it also allows, to some degree, negative correlations between responses within a litter. A likelihood ratio test is again proposed to test for differences between the experimental groups. For example, Kupper and Haseman demonstrated a better fit of their correlated binomial model than of the beta-binomial model. Recently, Paul (1985) proposed a three-parameter distribution that represents a generalization of the binomial, the beta-binomial, and the correlated binomial distributions.

All of the procedures proposed above are based on asymptotic approximation and rely on large m_i for their validity. This is particularly important when the response probabilities μ_i are near zero or one. Because many rodent lesions occur with frequencies of 1% or lower, exact permutation tests of the null hypothesis H_0: $\mu_1 = \mu_2$ against the alternative H_2: $\mu_1 > \mu_2$ based on the observed difference $d = \hat{p}_1 - \hat{p}_2$ may be considered (Crump and Howe, 1980). These tests are based on the fact that under the null hypothesis, each of the $s = (m_1 + m_2) C_{m1}$ possible assignments of m_1 litters to group 1 and m_2 litters to group 2 are equally likely. The significance level for the exact randomization test against the alternative H_1: $\mu_1 > \mu_2$ is then given by r/s, where r is the number of permutations leading to a value of d at least as large as the observed value. When s is large, Crump and Howe suggest that the significance level may be estimated on the basis of a random sample from the permutation distribution. In the special case

$$n_{ij} \equiv \mathbf{n},$$

the algorithm given by Soms (1977) may be used to obtain the significance level. Although these procedures are exact, they are conservative for small values of μ_1 in the sense that the false-positive rate can be notably less than the nominal level (Krewski et al., 1984).

7.3. Dose-Response Models

A number of statistical models have been developed to describe the dose-response relationships encountered in laboratory studies of carcinogenesis (Crump et al., 1977; Rai and Van Ryzin, 1981; Krewski and Van Ryzin, 1981; Gart et al., 1987). Although these models differ substantially with respect to their underlying biological rationale, they do not generally provide for the existence of intralitter correlation and, hence, are not directly applicable to situations of developmental toxicity. In this section, some recently proposed approaches to modeling dose-response curves for developmental toxicity data that can accommodate litter effects are reviewed.

Under the beta-binomial model considered by Williams (1975), the marginal distribution of x_{ij} (the number of animals responding in the jth litter in the ith treatment group) is beta-binomial, so that the likelihood for the parameters $\mu_i = \alpha_i/(\alpha_i + \beta_i)$ and $\pi_i = (\alpha_i + \beta_i)^{-1} > 0$ $(i = 1, \ldots, I)$ is a product of I independent beta-binomial terms. Segreti and Munson (1981) then proposed that the effects of the dose d_i administered to the ith group be modeled as

$$\mu_i = \lambda + (1 - \lambda)F(a + \beta \log d_i),$$

where $0 < \lambda < 1$, $\beta > 0$, and F is an (increasing) distribution such as the logistic. The beta-binomial likelihood may then be used to obtain estimates of the parameters α, β, and λ in the presence of dose-specific litter-effect parameters $\pi_1, \ldots, \pi_I$. (Segreti and Munson also consider a simpler but less realistic model in which $\pi_1 = \ldots = \pi_I = \pi$.) The former three estimates then provide a fitted dose-response curve

$$\mu = \lambda + (1 - \lambda)F(\alpha + \beta \log d).$$

Simulation studies using this model have shown that ignoring litter effects generally has no appreciable impact on the estimates of the logistic model parameters. However, not taking these effects into account can lead to significant underestimation of the variances associated with these parameter estimates (Kupper et al., 1986).

Williams (1982) introduced a class of weighted least-squares estimates for correlated binary regression estimation using only the beta-binomial mean and variance structure. These estimates are easily obtained using the GLIM computer program by modifying a standard linear logistic analysis to accommodate extrabinomial variation (see McCullagh and Nelder, 1983; Brooks, 1984). However, the standard errors may not be asymptotically correct under Williams's model (1982) because the asymptotic correlation between the estimates of α and β and the estimate of the intralitter correlation are not taken into consideration (Ochi and Prentice, 1984).

A different approach to this problem has been studied by Ochi and Prentice (1984). In general terms, they consider a correlated probit regression model in which the binary responses within the same litter are defined as indicators of whether or not the corresponding components of a multivariate normal regression vector with common mean and variance exceed some threshold value. Although the likelihood calculations are somewhat more complex than in the Segreti–Munson model, the Ochi–Prentice model provides for multiple covariates, such as litter size or maternal and pup body weights, as well as flexibility in modeling changes in intralitter correlation with dose.

Rai and Van Ryzin (1985) considered a dose-response model for teratological data in which the probability of a teratological response at a given dose varies with the litter size. Their procedure involves an exponential model for maternal toxicity that may lead to adverse effects on the individual pups. Under this model, the probability of response p_{ij} at dose d_i in a litter of size n_{ij} is given by

$$p_{ii} = [1 - \exp\{-(\alpha + d_i)\}] \exp\{-n_{ij} g(d_i)\},$$

where $\beta \geq 0$ and $g(d_i)$ is some function of the dose d_i. (Note that p_{ij} decreases as n_{ij} increases.) In particular, Rai and Van Ryzin assume g to be a linear function of dose d. The random variation in litter size is accommodated through the use of a Poisson distribution for the n_{ij} in which the expected litter size may decrease with dose.

Many of the previous procedures involve specific assumptions concerning intralitter correlation. Roberts et al. (1987) presented methods that do not require such assumptions. The basic input for these methods is the aggregate data $(x_{i\cdot}, n_{i\cdot})$ along with estimate of the variance of $p_i = x_{i\cdot}/n_{i\cdot}$ $(i = 1, \ldots, I)$. Suppose that the response probability p_i satisfies the logistic model

$$p_i = (1 + e^{\underset{\sim}{Z}_i \beta})^{-1},$$

where the $\underset{\sim}{Z}_i$ is a vector of covariates including dose and any other relevant variables, such as litter size, and $\underset{\sim}{\beta}$ is the corresponding vector of regression coefficients. A weighted least-squares estimator $\hat{\underset{\sim}{\beta}}_w$ of $\underset{\sim}{\beta}$ with weights based on empirical logits may then be obtained, along with the estimated variance-covariance matrix. The GLIM method may be applied to the aggregate data $(x_{i\cdot}, n_{i\cdot})$ to get a quasimaximum likelihood estimator $\hat{\underset{\sim}{\beta}}$ of $\underset{\sim}{\beta}$ (McCullogh and Nelder, 1983). Consistent estimates of the covariance matrix of $\hat{\underset{\sim}{\beta}}$ can also be obtained, along with goodness-of-fit tests and tests of nested hypotheses concerning the model parameters.

7.4. Summary and Conclusions

In this chapter, statistical methods for the analysis of developmental toxicity data were discussed. In particular, attention was focused on binary outcomes corresponding to the presence or absence of a particular anomaly noted in individual pups. Because observations are thus taken on littermates, the methods discussed are designed to accommodate any intralitter correlation that may be present. If such methods are not employed, it is possible that incorrect conclusions may be drawn, including serious overstatement of the statistical significance of any observed adverse effects.

Although a number of different statistical tests for comparing response rates in exposed and unexposed groups of test animals were discussed, no one method can be recommended to the exclusion of any others at this time. A difficulty with tests based on the beta-binomial model, which seems to have only recently been recognized, is a certain degree of instability in the maximum likelihood estimates of model parameters (Tamura and Young, 1986). Because of this, other estimates of these parameters warrant consideration (Tamura and Young, 1987). The exact randomization tests avoid this problem but can be somewhat conservative when the effect of interest occurs only infrequently in control animals. Most of these procedures also are applicable in testing for an increasing trend in response to increasing dose when data on two or more dose groups are available for analysis.

Individual dose-response models generally will require modification when applied to developmental toxicity data to provide for intralitter correlation. Such modifications are possible, as with the Segreti–Munson (1981) model in which the beta-binomial distribution is used in place of the binomial distribution in conjunction with a logistic or probit dose-response model. The correlated probit model proposed by Ochi and Prentice (1984) offers considerable flexibility in modeling both intralitter correlation and covariates other than dose but is comparatively somewhat complex. The model proposed by Rai and Van Ryzin (1985) has the unique feature of treating litter size as a random variable but allows the litter-specific response rates to vary primarily as a function of litter size. Simpler approaches can be invoked by working with the total number of effects noted within a treatment group, although this does not exploit the information on intralitter correlation available with data on individual litters.

References

Brooks, R.J. (1984) Approximate likelihood ratio tests in the analysis of beta-binomial data. *Appl. Stat.* 33:285–289.

Cochran, W.G. (1943) Analysis of variance for percentages based on unequal numbers. *J. Am. Stat. Assoc.* 38:287–301.

Crump, K.S. and Howe, R.B. (1980) A small-sample study of permutation tests for detecting teratogenic effects (Technical Report). Ebon Research Systems, Washington, D.C.

EPA (Environmental Protection Agency) (1985) Toxic Substances Control Act test guidelines; final rules. *Fed. Regist.* 50:39426–39428; 39433–39434.

FDA (Food and Drug Administration) (1970) Advisory committee on protocols for safety evaluations. Panel on Reproduction report on reproduction studies in the safety evaluation of food additives and pesticide residues. *Toxicol. Appl. Pharmacol.* 16:264–296.

Frangos, C.C. and Stone, M. (1984) On jackknife, cross-validity and classical methods of estimating a proportion with batches of different sizes. *Biometrika* 71:361–366.

Gad, S. and Weil, C.S. (1986) *Statistics and Experimental Design for Toxicologists*, The Telford Press, Caldwell, New Jersey.

Gart, J.J., Krewski, D., Lee, P.N., Tarone, R.E.D., and Washorendorf, F.J. (1987) Statistical Design in Analysis of *Long Term Animal Experiments*, vol. 3, International Agency for Research on Cancer, Lyon.

Gladen, B. (1979) The use of the jackknife to estimate proportions from toxicological data in the presence of litter effects. *J. Am. Stat. Assoc.* 74:278–283.

Haseman, J.K. and Hoel, D.G. (1974) Tables of Gehan's generalized Wilcoxon test with fixed point censoring. *J. Statistical Computation Simulation* 3:117–135.

Haseman, J.K. and Hogan, M.D. (1975) Selection of the experimental unit in teratology studies. *Teratology* 12:165–172.

Haseman, J.K. and Kupper, L.L. (1979) Analysis of dichotomous response data from certain toxicological experiments. *Biometrics* 35:281–293.

Haseman, J.K. and Soares, E.R. (1976) The distribution of fetal death in control mice and its implications on statistical tests for dominant lethal effects. *Mutat. Res.* 41:277–288.

Hogan, M.D. and Hoel, D.A. (1982) Extrapolation to man. In *Principles and Methods of Toxicology* (Hayes, Ed.), pp. 711–731, Raven Press, New York.

Hollander, M. and Wolfe, D.A. (1973) *Nonparametric Statistical Methods*, John Wiley & Sons, New York.

Kleinman, J.C. (1973) Proportions with extraneous variance: Single and independent samples. *J. Am. Stat. Assoc.* 68:46–54.

Krewski, D., Brennan, J., and Bickis, M. (1984) The power of the Fisher permutation test in 2 x k tables. *Commun. Stat. Simul. Com.* B13:433–448.

Krewski, D. and Van Ryzin, J. (1981) Dose response models for quantal response toxicity data. In *Statistics and Related Topics* (M. Csorgo, D.A. Dawson, J.N.K. Rao, and A.K.M.E. Saleh, Eds.), pp. 201–231, North Holland, New York.

Kupper, L.L. and Haseman, J.K. (1978) The use of a correlated binomial model for the analysis of certain toxicological experiments. *Biometrics* 34:69–76.

Kupper, L.L., Portier, C., Hogan, M.D., and Yamamoto, E. (1986) The impact of litter effects on dose-response modeling in teratology. *Biometrics* 42:85–98.

Lin, F.O. and Haseman, J.K. (1976) A modified Jonckheere test against ordered alternatives when ties are present at a single extreme value. *Biometrische Zeit.* 18:623–631.

Mantel, N. (1969) Some statistical viewpoints in the study of carcinogenesis. In *Progress in Experimental Tumor Research*, Vol. 11, pp. 431–443, S. Karger, New York.

McCullagh, P. and Nelder, J.A. (1983) *Generalized Linear Models*, Chapman & Hall, London.

Ochi, Y. and Prentice, R.L. (1984) Likelihood inference in a correlated probit regression model. *Biometrika* 71:531–543.

Oser, B.L. and Oser, M. (1956) Nutritional studies on rats on diets containing high levels of partial ester emulsifiers. II. Reproduction and lactation. *J. Nutr.* 60:489–505.

Paul, S.R. (1985) A three-parameter generalization of the binomial distribution. *Commun. Statist. Theor. Meth.* 14:1497–1506.

Prentice, R.L. (1986) Binary regression analysis using an extended beta-bionomial distribution with discussion on correlation induced by co-variate measurement errors. *J. Am. Stat. Assoc.* 81:321–327.

Rai, K. and Van Ryzin, J. (1981) A generalized multi-dose response model for low-dose extrapolation. *Biometrics* 37:341–352.

Rai, K. and Van Ryzin, J. (1985) A dose-response model for teratological experiments involving quantal responses. *Biometrics* 41:1–9.

Rao, J.N.K. and Scott, A.J. (1981) The analysis of categorical data from complex sample surveys. Chi-squared tests for goodness of fit and independence in two-way tables. *J. Am. Stat. Assoc.* 76:221–230.

Roberts, G., Rao, J.N.K. and Kumar, S. (1987) Logistic regression analysis of sample survey data. *Biometrika* 74:1–12.

Schreiner, C.A. and Steelman, J.R. (1977) Oral triethylene-melamine: Effects in the dominant lethal assay for mutagenicity. *Toxicol. Appl. Pharmacol.* 42:487–495.

Segreti, A.C. and Munson, A.E. (1981) Estimation of the median lethal dose when responses within a litter are correlated. *Biometrics* 37:153–156.

Shirley, E.A.C. and Hickling, R. (1981) An evaluation of some statistical methods for analyzing numbers of abnormalities found amongst litters in teratology studies. *Biometrics* 37: 819–829.

Soms, A.P. (1977) An algorithm for the discrete Fischer's permutation test. *J. Am. Stat. Assoc.* 72:662–664.

Southward, G.M. and Van Ryzin, J. (1972) Estimating the mean of a random binomial parameter. In *Proceedings of the Sixth Berkeley Symposium IV* (Le Cam, J. Neyman, and E.L. Scott, Eds.), pp. 249–263, University of California Press, Berkeley.

Stucky, W. and Vollmar, J. (1976) Exact probabilities for ties linear rank tests. *J. Statistical Computation Simulation* 5:73–81.

Tamura, R. and Young, S. (1986) The incorporation of historical control information in tests of proportions: Simulation study of Tarone's procedure. *Biometrics* 42:343–349.

Tamura, R. and Young, S. (1987) A stabilized moment estimator for the beta-binomial distribution. *Biometrics* 43:813–824.

Weil, C.S. (1970) Selection of the valid number of sampling units and a consideration of their combination in toxicological studies involving reproduction, teratogenesis or carcinogenesis. *Food Cosmet. Toxicol.* 8:177–182.

Williams, D.A. (1975) The analysis of binary responses from toxicological experiments involving reproduction and teratogenicity. *Biometrics* 31:949–952.

Williams, D.A. (1982) Extra-binomial variation in logistic linear models. *Appl. Stat.* 31:144–148.

Chapter VIII

Epidemiology

8.1. Introduction

Over the last 25 years the number of women employed outside the home has risen dramatically in many countries. For example, in Canada the number has risen from 1.8 million in 1962 (29% of all women 15 years of age and older) to 4.9 million (52%) in 1982 (Statistics Canada, 1983). This increased participation rate—the number of women in the labor force as a percentage of all women in the population—has been mirrored by a trend for women to continue working for a greater proportion of their pregnancies. A recent survey (Statistics Canada, 1982) estimated that in 1981, 150,000 women were absent from work for two or more weeks during the year because of pregnancy. Although this number is only an approximation of the total number of women in the labor force who were pregnant during that year, it is an indication of the large number of pregnancies potentially at risk from the occupational environment.

The realization that environmental factors may present risks for reproductive health began about the turn of the century when excesses of stillbirth, spontaneous abortion, and neonatal death after occupational exposure to high levels of lead were reported in several European countries and Japan (Rom, 1980). The teratogenic effects of ionizing radiation were reported in 1920, those of the rubella virus in 1941, those of aminopterin in 1952, and those of thalidomide in 1961 (Kalter and Warkany, 1983). It was the thalidomide discovery that accelerated research and increased concern about the effects of prenatal exposure to chemicals and possible fetal damage. The

Table 8-1. Chemicals Presenting a Risk to the Developing Fetus Based on Reliable Epidemiological and Teratological Evidence

Chemical	Adverse reproductive effects	Reference
Anaesthetic gases	Spontaneous abortion, congenital malformations(?)	Spence and Knill-Jones, 1978; Vessey and Nunn, 1980
Ethylene oxide	Spontaneous abortion	Hemminki et al., 1982
Styrene	Spontaneous abortion, toxemia, congenital malformations	Hemminki et al., 1980; Holmberg, 1979; Murray et al., 1978
PCBs (yusho, oil disease)	Toxemia, low birth weight, stillbirth, missed abortion	Barlow and Sullivan, 1982; Bercovici et al., 1983; Wassermann et al., 1982
Inorganic lead	Stillbirth, miscarriage, premature membrane rupture	World Health Organization, 1977; Fahim et al., 1976
DDT	Premature birth	O'Leary et al., 1971
Carbon monoxide	Fetal and infant death, microcephaly, mental retardation	Barlow and Sullivan, 1982
DES	Vaginal cancer in female offspring, urological anomalies in males	Herbst et al., 1971; Gill et al., 1977
Inorganic and organic mercury	Congenital cerebral palsy, prenatal mortality	Barlow and Sullivan, 1982; Tsubaki and Irukayama, 1977

role of environmental chemicals in reproduction was reinforced when the effects of methyl mercury and polychlorinated biphenyls became known (Koos and Longo, 1987; WHO, 1976).

The increasing number of chemicals in the occupational environment has caused concern over their possible adverse effects on reproduction because exposure to many chemicals tends to be higher in the workplace than in the general environment. However, systematic investigations of the risks for the reproductive health of workers and their families before, during, and after pregnancy are few, particularly for female-dominated occupations and for females in male-dominated occupations (Heinonen, 1986; Stellman, 1978).

Very little reliable epidemiological (i.e., human) evidence of adverse health effects on the mother and/or fetus exists for specific chemicals encountered in the workplace (Table 8-1). Some additional limited epidemiological and toxicological evidence is suggestive of possible adverse reproductive effects after exposure to a small number of other chemicals (Table 8-2). A relatively large

Table 8-2. Chemicals Presenting a Risk to the Developing Fetus Based on Suggestive Epidemiological and Teratological Evidence

Chemical	Adverse reproductive effects	Reference
Benzene	Stillbirth, spontaneous abortion, aplastic anemia, perinatal death	Barlow and Sullivan, 1982
Vinyl chloride	Congenital malformations of various kinds	Infante, 1979; Barlow and Sullivan, 1982
2,4-D	Spontaneous abortion, premature birth, toxicosis	Carmelli et al., 1981
Arsenic	Infant death, pulmonary hemorrhage, low birth weight	Nordstrom et al., 1979; Barlow and Sullivan, 1982
Carbon disulfide	Spontaneous abortion	Hemminki et al., 1980

number of epidemiological studies have linked employment in certain occupational groups or industries with an increased risk of adverse reproductive outcome (usually spontaneous abortion or congenital malformation and less often perinatal death, prematurity, and childhood tumors [Axelson et al., 1983; Blomqvist et al., 1981; Shilling and Lalich, 1984]). For a comprehensive recent review, see Lindbohm et al. (1985). However, this latter type of study cannot be considered definitive, but rather a hypothesis-generating exercise for isolating causal mechanisms. The ways in which epidemiologic and toxicologic methods may be used to study these cause-effect relationships are the basis of the remainder of this chapter.

8.2. The Conduct of Epidemiologic Studies on Human Reproduction

The basic requirement of an epidemiologic study of reproductive failure in relation to environmental exposures is that exposure and outcome must be selected and defined in advance. The outcome of interest is usually a list of undesirable endpoints (several common endpoints are shown in Table 8-3). Many of these endpoints are not independent, e.g., neonatal mortality rates may increase as late fetal death decreases. However, this shortcoming is usually overlooked (Edmonds et al., 1981).

The definition of exposure and how it can be quantified is usually difficult. These difficulties will be discussed in more detail in the next section. For the time being, the implication is simply that an adverse exposure occurs when

Table 8-3. Frequency of Selected Reproductive Failure Outcomes[a]

Event	Occurrence/100 births	Unit of study
Low birth weight (< 2500 g)	1	Livebirths
Spontaneous abortion (< 29 weeks gestation)	10–20	Pregnancies
Chromosomal anomaly in spontaneous abortuses	30–40	Spontaneous abortions
Stillbirth	1	Total births
Birth defects	3–6	Total births
Severe mental retardation	0.4	Children to age 15 years

[a] Adapted from Bloom, A.D., Ed. (1981) *Guidelines for Studies of Human Populations Exposed to Mutagenic and Reproductive Hazards.* March of Dimes Birth Defects Foundation, White Plains, New York, p. 47, with permission.

individuals come into contact with a contaminant in the physical environment concentrated in specific areas, such as the workplace.

Given that an exposure and an outcome can be identified, the epidemiologic study that follows can be conducted at either the understanding or the intervention level (Kleinbaum et al., 1982). At the understanding level, research proceeds from observation to inference about the natural history of the disease and its etiology. At the intervention level, information is collected that can be used to prevent, eradicate, or control the disease. Because the epidemiology of reproductive disorders as related to environmental chemical exposure is not well developed, most of what follows relates to strategies for observational research. Observational studies are usually divided into two categories: descriptive and etiologic studies.

8.3. Descriptive Studies

A descriptive (ecologic) study is often the design of choice when little is known about the causes, natural history, occurrence, or population at risk. The main objectives are to determine the frequency of the outcome in the population under study, to search for correlations or associations of disease occurrence with easily accessible covariate information, and thus to generate more specific etiologic hypotheses.

Descriptive studies are usually carried out quickly and inexpensively with available information. They seldom involve purposely collected data on individuals. This requirement naturally restricts the choices of reproductive endpoint, population at risk, and exposure measurement. The information on exposure is usually the most seriously limited.

Typically the first step in a descriptive study is to select the outcomes of interest. Valid, long-term, population-based information on late fetal death, low birth weight or premature birth, congenital birth defects, and infant death are readily available in many countries and are commonly used as indices of reproductive failure. Once the endpoints have been identified—descriptive studies often use many endpoints—the geographic area and the population at risk (and control populations, if used) must be well defined. A population-based study is essential to calculate a denominator for comparison purposes.

At this point it is necessary to assess the quality of the endpoint data by verifying diagnoses (if possible) and identifying possible problems with data collection or reporting systems that might result in biased ascertainment or misclassification. Finally, covariate and exposure information is identified and collected on the same geographical basis.

The analysis of a descriptive study is usually straightforward and takes one of two forms: a simple comparison of frequency or rate between two or more groups at one time or during a period of time, or an assessment of trend in one or more groups over time. These analyses can be combined to estimate the change in average exposure level and the corresponding change in disease occurrence for several groups (Kleinbaum et al., 1982).

8.4. Etiologic Studies

Etiologic (analytic) studies are conducted when sufficient information about the disease is known so that specific a priori hypotheses can be tested. This enables the identification of risk factors, the estimation of their effects on the disease, both alone and in combination with other agents and the development of possible prevention or intervention strategies.

Etiologic studies are invariably more time consuming and expensive than descriptive studies, but they are indicated when a serious problem exists that is not amenable to study with a descriptive approach or when a descriptive study has demonstrated a statistically significant association so that it is necessary to specify more closely the estimate of risk or exposure-response relationship.

There are two basic types of etiologic studies: the cohort (prospective) and case-control (retrospective) approaches. These approaches have some similarities and many differences in basic design and operational options (e.g., directionality, timing, subject selection, sampling strategy, and units of observation and analysis). In almost all cases, however, the unit of observation in etiologic studies is the individual, whereas the unit in descriptive studies is aggregates of people.

In a cohort study, information on exposure is known for all subjects at the beginning of the follow-up period. The population under study is then followed for a given period during or after which individuals developing the outcome of interest are identified. Cohort studies can take place in present time, i.e., exposure status is determined in the present, and follow-up is started in the future, but more commonly exposure is determined in the past, (e.g., from occupational records), and follow-up proceeds in the present if the interval is sufficient.

Case-control studies involve a nondirectional or backward design that compares a group of cases (i.e., individuals known to have the endpoint of interest) with one or more groups of noncases with respect to a current or previous exposure. Study subjects are selected from separate populations of cases and noncases, and exposure status is then determined, often by direct (interview) contact with each subject in the study.

The cost of linking available information from various sources and collecting new information from individuals to assemble data of the amount and specificity needed to justify an etiologic study is an important consideration. The cost of the study should be compared with the expected return, which is often expressed in terms of statistical power, i.e., the ability to detect and estimate true differences in occurrence. Power depends in a complex way on the usual rate of the outcome of interest in the unexposed population, the level of significance that is accepted as confirmation of an association (often 1 in 20 or $p = 0.05$), the sample sizes of exposed and unexposed subjects available to the study, and the true (but unknown) magnitude of the expected effects. The expected statistical power of an etiologic study should be carefully estimated early in the planning phases.

All etiologic study designs are susceptible to distortions in the estimation of effect, also called bias (Kleinbaum et al., 1982). At least 40 sources of bias have been identified for case-control studies (Sackett, 1979), and most of them can apply to cohort studies. Bias is generally difficult to quantify precisely. It may only be possible to determine the direction of a bias, but this can be of great importance, e.g., if it can be shown that a weak association is probably stronger than it appears because of bias. Bias is minimized by sound study design, a sophisticated analysis, and replication studies.

Reproductive disorder studies, whether prospective or retrospective, possess certain unique characteristics that have implications in their design and conduct. One of the most important of these is the time interval between exposure and outcome. Although relevant exposure may have occurred long in the past, for many adverse outcomes it is possible to limit the period of exposure to the duration of the pregnancy (or a little more)—a period much shorter than is necessary for the study of most chronic diseases. This short latent period will tend to reduce certain types of bias (e.g., recall bias), reduce the number of study dropouts, and make access to supporting documents (e.g., medical and occupational records) more likely.

8.5. Characteristics of an Acceptable Human Reproduction Study

Determining if a study is acceptable for risk assessment purposes follows naturally from the way the study was designed and conducted. Observational studies and their variants are most useful when large (but relatively sparse) data sources can be combined (e.g., the Labor Force Surveys and vital statistics) to obtain information on both the study factors and diseases in the same populations. Care must be taken not to apply observational associations from one level of aggregation (e.g., a health unit or county) to a different level (e.g., an individual) because the magnitude of association between two sociodemographic variables tends to be much higher in observation studies than in individual analyses (Stavraky, 1976).

Although observational studies have many limitations with respect to data availability and the inability to test hypotheses, they are useful for generating hypotheses and for evaluating the efficacy and effectiveness of intervention programs.

Cohort designs are best suited to study relatively frequent outcomes. Prevalent cases at the beginning of the follow-up period should be excluded, and all study subjects should be followed for the same time period in the same way. A potential problem in any type of cohort study is the loss of subjects because of migration, nonparticipation, and death (if death is not an endpoint). Excessive attrition of the study population can lead to distortion of results that cannot be easily corrected for in the analysis (Kleinbaum et al., 1982).

Cohort studies have the major advantage of being relatively free of selection bias because the exposure status of each subject is determined before the disease is detected. Because exposure and other covariate information must be decided at the beginning of the study, however, this design is not able to generate or test new hypotheses after follow-up has begun.

Case-control studies are particularly well suited to hypothesis testing on specific rare diseases. In general, case-control studies are less time consuming and less expensive than cohort studies because of smaller sample sizes required, relatively brief study duration, and a choice of convenient sampling strategies. Because the ratio of cases to noncases can be fixed before the study begins, analyses are more statistically efficient than for other designs of the same size.

Case-control studies are primarily limited by the fact that it is difficult to ensure that cases and noncases in the study are comparable with respect to possible confounding and effect-modifying factors because they are selected from two separate populations. Other practical drawbacks of case-control studies include a susceptibility to subject recall bias, which usually operates to make the observed association appear stronger than the true association.

The two most important general considerations for acceptability of etiologic studies are validity (i.e., freedom from bias) and statistical power. Statistical power is based on design considerations and a priori knowledge of the disease process. A brilliantly conducted study can be completely negated by the fact that an inadequate sample size, for example, guaranteed an inability to distinguish between a real and a chance association. Validity is based on both design and statistical considerations. It is considered to be the sine qua non of etiologic research and is more important than precision (lack of random error) in risk estimation (Kleinbaum et al., 1982).

In general, prospective cohort studies involving incident cases over a short follow-up period might be considered the optimal design for testing etiologic hypotheses, whereas nondirectional designs using secondary data sources would be considered much less informative. Usually, however, a multitude of practical difficulties conspire to make the ideal study infeasible. As a result, concessions are made, increasing the number of assumptions necessary to support a causal interpretation. Thus, one of the most important qualities of an acceptable study is that its key assumptions are made explicit so that the reader can judge their importance (Bross, 1979).

A large body of evidence suggests that the fetus is sensitive to a wide variety of environmental exposures throughout the gestational period. However, this evidence is either incomplete or altogether lacking for many exposure settings. Occupational exposures of potentially high risk are not, in general, a matter of personal choice, and studies are needed to quantify the risks to the fetus with an aim to limiting potential adverse sequelae by transferring workers to safer jobs or by lowering workplace exposure levels. The best available method to achieve this aim appears to be with a synthesis of toxicological exposure and endpoint determination and epidemiological sampling and analysis techniques.

References

Axelson, O., Edling, C., and Andersson, L. (1983) Pregnancy outcome among women in a Swedish rubber plant. *Scand. J. Work Environ. Health* 9 (suppl. 2):79–83.

Barlow, S.M., and Sullivan, F.M. (1982) *Reproductive Hazards of Industrial Chemicals: An Evaluation of Animal and Human Data*, Academic Press, London.

Bercovici, B., Wassermann, M., Cucos, S., Ron, M., Wassermann, D., and Pines, A. (1983) Serum levels of polychlorinated biphenyls and some organochlorinated insecticides in women with recent and former missed abortions. *Environ. Res.* 28:169–174.

Blomqvist, U., Ericson, A., Kallen, B., and Westerholm, P. (1981) Delivery outcome for women working in the pulp and paper industry. *Scand. J. Work Environ. Health* 7:114–118.

Bross, I.D.J. (1979) Right answers from wrong assumptions. *Prev. Med.* 5:203–206, editorial.

Carmelli, D., Hofherr, J., Tomsic J., and Morgan, R.W. (1981) *A Case-Control Study of the Relationship between Exposure to 2,4-D and Spontaneous Abortions in Humans*, SRI International.

Edmonds, L., Shrout, P., Hatch, N., Stein, Z., Holmes, L., Warburton, D., Klein, J., Weinstock, M., Letz, G., Whorton, R., Levin, B., Wyrocek, A., and Miller, R. (1981) Guidelines for reproductive studies in exposed human populations. In *Guidelines for Studies of Human Populations Exposed to Mutagenic and Reproductive Hazards* (A.D. Bloom, Ed.), pp. 39–110, March of Dimes Birth Defects Foundation, White Plains, New York.

Fahim, M.S., Fahim, Z., and Hall, D.G. (1976) Effects of subtoxic lead levels on pregnant women in the state of Missouri. *Res. Commun. Chem. Pathol. Pharmacol.* 13:309–331.

Gill, W.B., Schumacher, G.F.B., and Bibo, M. (1977) Pathological semen and anatomical abnormalities of the genital tract in human male subjects exposed to diethylstilbestrol in utero. *J. Urol.* 117:477–480.

Heinonen, O.P. (1986) Spontaneous abortions, stillbirths and birth defects in epidemiological search for risk indicators. In *New Concepts and Developments in Toxicology* (P.L. Chambers, P. Gehring, and F. Saki, Eds.), pp. 63–76, Elsevier Science Publishers, New York.

Hemminki, K., Franssila, E., and Vainio, H. (1980) Spontaneous abortions among female chemical workers in Finland. *Int. Arch. Occup. Environ. Health* 45:123–126.

Hemminki, K., Mutanen, P., Saloniemi, I., Niema, M.L., and Vainio H. (1982) Spontaneous abortions in hospital staff engaged in sterilizing instruments with chemical agents. *Br. Med. J.* 285:1461–1463.

Herbst, A.L., Ulfelder, H., and Poskanzer, D.C. (1971) Adenocarcinoma of the vagina: Association of maternal stilbestrol therapy with tumor appearance in young women. *N. Engl. J. Med.* 284:878–881.

Holmberg, P.C. (1979) Central-nervous-system defects in children born to mothers exposed to organic solvents during pregnancy. *Lancet* ii:177–179.

Infante, P.F. (1979) Oncogenic and mutagenic risks in communities with polyvinyl chloride production facilities. *Ann. N.Y. Acad. Sci.* 271:49–57.

Kalter, H., and Warkany, J. (1983) Congenital malformations: Etiologic factors and their role in prevention. *N. Engl. J. Med.* 308:424–431.

Kleinbaum, D.G., Kupper, L.L., and Morgenstern, H. (1982) *Epidemiologic Research*, Lifetime Learning Publications, Belmont, California.

Koos, B.J., and Longo, D.L. (1987) Mercury toxicity in the pregnant woman, fetus, and newborn infant: A review. *Am. J. Obstet. Gynecol.* 126:390–409.

Lindbohm, M-L., Taskinen, H., and Hemminki, K. (1985) Reproductive health of working women: Spontaneous abortions and congential malformations. *Public Health Rev.* 13:55–87.

Murray, F.J., John, J.A., Balmer, M.F., and Schwetz, B.A. (1978) Teratologic evaluation of styrene given to rats and rabbits by inhalation or by gavage. *Toxicology* 11:335–343.

Nordstrom, S., Beckman, L., and Nordenson, I. (1979) Occupational and environmental risks in and around a smelter in northern Sweden, V. Spontaneous abortion among female employees and decreased birth weight in their offspring. *Hereditas* 90:291–296.

O'Leary, J.A., Davies, J.E., Edmundson, W.F., and Feldman, M. (1972) Correlation of prematurity and DDE levels in fetal whole blood. In *Epidemiology of DDT* (J.F. Davies and W.F. Edmundsen, Eds.), pp. 55–56, Futura Publishing, Mount Kisco, New York.

Rom, W.N. (1980) Effect of lead on reproduction. In proceedings of a Workshop on Methodology for Assessing Reproductive Hazards in the Workplace (P.F. Infante and M.S. Legator, Eds.), DHHS, NIOSH Publication No. 81-100, pp. 33–42, U.S. Department of Health and Human Services, Washington, D.C.

Sackett, D.L. (1979) Bias in analytic research. *J. Chronic Dis.* 32:51–63.

Shilling, S., and Lalich, N.R. (1984) Maternal occupation and industry and the pregnancy outcome of U.S. married women, 1980. *Public Health Rep.* 99:152–161.

Spence, A.A., and Knill-Jones, R.P. (1978) Is there a health hazard in anaesthetic practice? *Br. J. Anaesth.* 50:713–719.

Statistics Canada (1982) *The Labour Forces, November 1982*, Statistics Canada Cat. No. 71-001, Ottawa.

Statistics Canada (1983) *Labor Force Occupation Trends*, Statistics Canada Cat. No. 92-920, Ottawa.

Stavraky, K.M. (1976) The role of ecologic analysis in studies of the etiology of disease: A discussion with reference to large bowel cancer. *J. Chronic Dis.* 29:435–444.

Stellman, J.M. (1978) Occupational health hazards of women: An overview. *Prev. Med.* 7:281–293.

Tsubaki, T., and Irukayama, K. (1977) *Methylmercury Poisoning in Minamata and Niigata, Japan.* Elsevier Press, Amsterdam.

Vessey, M.P., and Nunn, J.F. (1980) Occupational hazards of anaethesia. *Br. Med. J.* 281:696–698.

Wassermann, M., Ron, M., Bercovici, B., Wassermann, D., Cucos, S., and Pines, A. (1982) Premature delivery and organochlorine compounds: Polychlorinated biphenyls and some organochlorine insecticides. *Environ. Res.* 28:106–112.

WHO (World Health Organization) (1976) *Polychlorinated Biphenyls and Terphenyls*, Environmental Health Criteria 2, World Health Organization, Geneva.

WHO (World Health Organization) (1977) *Lead*, Environmental Health Criteria 2, U.N. Environment Program and World Health Organization, Geneva.

Chapter IX

Extrapolation of Reproductive Risks From Animal and Human Data

9.1. Introduction

Assessment of human reproductive risk from developmental toxicity data involves three steps: (1) conducting animal and, if possible, human studies; (2) evaluating these data; and (3) using these data in the extrapolation of reproductive risk to humans. The last step is derived from toxicology data and is based on a system of concepts and conservative suppositions. Some of these concepts, such as maternal health and its possible influence on fetal and postnatal development as well as the relative significance of fetal aberrations and malformations, have recently received considerable attention regarding their importance in human risk estimation. Based on these revised concepts and other newer ones that are logically admissible, a framework of risk extrapolation is proposed in this chapter.

Historically, developmental toxicity was ushered in by concerns about malformations in humans. Subsequently, many studies designed to examine the teratogenic potential of chemicals in animals were conducted. As a consequence the bulk of data dealing with developmental toxicity pertains to teratologic effects, which are considered in greater detail in this chapter. A similar comprehensive evaluation of other endpoints of developmental toxicity is not possible at present because the relevant data are not available. Despite this consideration, it is important to examine the total developmental toxic response of the conceptus to a chemical and not malformation alone.

The judicious extrapolation of human risks from reproductive studies requires a comprehensive database, a broad base of knowledge, and familiarity with the performance of complex studies. In attempting to extrapolate from data obtained in such studies to the human situation, consideration must be given not only to the differences but also to the similarities of the reproductive process in humans and animal test species. A reproductive study should be interpreted only by an experienced individual with a broad basic knowledge of the toxicology and metabolism of the test material.

Ethical considerations preclude testing new chemicals or suspected hazardous chemicals in humans. Thus, regulatory risk assessment consists of weighing evidence from in vitro and in vivo tests. In addition to the considerable difficulties in evaluating evidence from a number of these tests, there is an additional, and perhaps more questionable, assumption that the results are predictive for the human population. Therefore, data from epidemiological studies are extremely valuable when available to provide confirmation of the degree of correlation between animal and human susceptibility.

The basic toxicity studies that are conducted today have not changed fundamentally in the last 25 years. They are enlarged in scope, use refined measurement and analysis, and cost a great deal more, but they still too often provide a package of conflicting or incomplete evidence that the regulator must use as the basis for a risk-benefit decision. To use an example from teratology, in 1976 Staples reported the figure of 800 chemicals identified as teratogens in laboratory animals, but fewer than 25 are known human teratogens. The situation has not changed since then, except that the ratio of animal to human teratogens has increased (Larsson et al., 1982). When the correlation between known effects in laboratory animals and clinical adverse effects in humans is so low, the value of routine tests in animals for predicting teratogenesis must be questioned. In part, however, this discrepancy may arise from the high loss rates reported in human pregnancies, and such a loss may result in early elimination of aberrant products of conception. Data are not available, however, to confirm or refute this possibility. In addition, dose levels found to be teratogenic in animal studies are often much higher than the human therapeutic doses, and the observed teratogenicity may be associated only with maternotoxic dose levels.

In most cases data from animal studies are insufficient to permit a precise estimate of human risk because of the difficulties of transspecies extrapolation. The test batteries are useful in identifying toxic compounds that cause catastrophic events, such as carcinogenesis, teratogenesis, or fetal death, especially if they do so in a variety of test systems at or near doses actually encountered by humans. However, when only subtle or minimal changes are induced in only some tests or only at high doses or when the mechanism of toxicity is not defined (as is usually the case with the human embryo-fetus, for example), the value of the tests as an indicator or a potential reproductive problem in humans is considerably diminished.

There has been an increasing pressure to include testing for behavioral and organ functions of progeny prenatally exposed to chemicals in the scheme of risk estimation. It is also being strongly advocated that in vitro tests should be used as a preliminary step in screening chemicals in whole-animal studies. To date, behavioral studies, while often indicating effects, have not proved to be as sensitive as traditional observations of reproductive studies and are therefore not a major factor in estimating potential human risk. Organ function studies may detect indications of biochemical abnormalities, but frequently these are noted in ancillary studies performed for other purposes. The state of the art with respect to in vitro studies is currently limited to screening of compounds. All three systems need further refinement and validation before they can be used effectively and meaningfully in assessment of human risk.

9.2. Classification of Chemicals for In Utero Effects

While estimating safe human exposure levels, an evaluator frequently determines uncertainty (safety) factors, the magnitude of which vary depending on the type, severity, and frequency of observed developmental toxic effects. The basic indicators that have been used to evaluate reproductive toxicity effects are identical for all chemicals and include the following: implantation losses, early and late resorptions, abortions and stillbirths, embryo-fetal growth (weight and length), and malformations. However, the weighting of these indicators now is frequently modified by ancillary data (e.g., pharmacokinetics). A basic problem with the system is an inability to utilize underlying mechanisms of action because of a lack of knowledge of such mechanisms for individual chemicals. The discussion that follows draws examples primarily from prenatal studies, but the principles outlined apply equally to multigeneration and fertility studies.

From experimental results, chemicals or drugs can be divided into four general categories rated from the highest bracket, A, to the lowest bracket, D, depending on an arbitrary degree of potential risk for humans. Potential risk is defined as the capacity of a chemical to induce fetal malformations, embryo-fetal death, and postnatal effects in humans. This capacity is not evident at the time of extrapolation to the human situation, and it may or may not become apparent in the future.

The chemical compounds that have so far been incriminated for inducing birth defects in humans after their intake during pregnancy (Wilson, 1973; Schardein, 1976; Strobino et al., 1978; and Frankos, 1985) are also similarly divided into four categories, A through D. The main characteristics of these categories are summarized in Table 9-1.

The classification is based primarily on malformations, survival, and growth during embryo-fetal development; data on prenatal effects of chemicals in humans and animals are broad enough for its justification. However,

Table 9-1. Arbitrary Classification of Chemicals Based on Teratogenic Potential

Criteria	Category A	Category B	Category C	Category D
1. Ratio: minimum maternotoxic dose to minimum teratogenic dose	Much greater than 1	Generally greater than 1, teratogenic range starts below the maternotoxic dose dose-range[a] and overlaps it	≤ 1	No teratogenicity even at maternotoxic doses
2. Incidence of malformations	Dose related and high	Dose related and high	Dose relatedness of each malformation less obvious, incidence low	–
3. Type of malformation at lower doses	Organ systems involved are specific	Characteristics, possibly specific, generally multiple	Nonspecific involving different organ systems	–
4. Target cells	Specific cells	Specific cells	Nonspecific and generalized	Not known
5. Range of safety factor	1–400	1–300	1–250	1–100

[a]The maternotoxic dose range extends between the dose initiating signs of toxicity and the dose causing 50% mortality (LD_{50}).

such a classification seems equally valid for postnatal effects of chemicals after prenatal exposure. At the present time this proposed classification is tentative and has not been adopted by any regulatory agency.

9.2.1. Category A

Chemicals tested in experimental animals that fall into this category are selectively active on the fetus, causing irreversible and characteristic morphologic fetal alterations or embryo-fetal deaths at doses well below the maternotoxic dose range. The type of fetal alteration and the affected organ system are generally characteristic of the chemical because their effects appear to be selective for specific fetal tissues (Table 9-2). The lesions or malformations may be few and specific at low doses, but multiple and diverse at high dose levels. In humans the compounds in this category (Table 9-3) seem to cause somewhat specific malformations that are generally reproducible in one or more experimental mammalian species at apparently nonmaternotoxic dosages. There is no documented evidence that these human teratogens (Table 9-3) cause fetal wastage (embryo-fetal death, abortion, or stillbirth) at the usual therapeutic doses. However, detection of an increase in conceptus wastage would be difficult in light of the normally high background loss rate.

An appropriate subject for further research is the concept of the existence of a primary site of action or target cells with the possible presence of cellular receptors specific for a true, or selective, teratogen as an initial step in dysmorphogenesis. The receptor concept and the importance of the ligand interaction in altering cellular responses have been reviewed (Kimmel, 1981). For elucidating chemical-specific target cells, minimum effective doses should be investigated because heavy doses may overshadow the specific target cells by also causing lesions of cells in multiple organ systems.

9.2.2. Category B

The chemicals belonging to category B produce a high, dose-related incidence of reproductive toxic effects (malformation or embryo-fetal death) in mammalian species at doses that may or may not be maternotoxic (Table 9-4) and a low incidence of the adverse fetal effects at nonmaternotoxic doses.

The malformations observed at maternotoxic doses can be characterized by two components: (1) an embryonic or specific component, which probably results from a chemical's direct effect on the embryonic tissues (Table 9-4) and (2) a maternal or nonspecific component probably resulting from a chemical's toxic effect on the maternal organism. In cases of fetal malformations induced at nonmaternotoxic doses of chemicals, no maternal component may be involved.

Table 9-2. Category A: A Probable True or Selective Action in Animals

Chemical	Minimum effective dose[a] (mg/kg), route, and days of dosing during pregnancy	Maternal toxicity of the minimum effective dose	Maternotoxic dose, acute, species	LD_{50} (mg/kg)[b]	Salient malformations and references
Mercaptopurine riboside	62.5 ip, 11	Not reported	Rat	2,000–3,000	Cleft palate; adactyly, ectrodactyly, syndactyly, or brachydactyly; short tail; kidneys absent, hypoplastic, or fused; adrenal extopic or absent (Kury et al., 1968)
Azaserine	2.5 ip; 11, 12, or 13	Not reported	Rat	75–100	Cleft palate; fused ribs and vertebrae; hemivertebrae; fused sternebrae; small or absent pelvis, femur, and fibula; syndactyly (Murphy and Karnofsky, 1956; Murphy et al., 1957)
$Ni(CO)_4$	0.08 mg per liter inhalation for 15 minutes	Not reported	Rat	0.58 mg/L for 15 min.	Anophthalmia and microphthalmia (Sunderman et al., 1979)
Medroxy progesterone acetate	3 or 10 sc, 10–12	None	Rabbit		Cleft palate (Andrew and Staples, 1977)
Ethylenethiorea	30 po, 13	None	Rat	60	Hydrocephalus and microphthalmia (Khera, 1973; Khera and Tryphonas, 1977)
Mirex	5 or 6 po, 6–15 or 8½–15½	None	Rat	Not reported	Subcutaneous edema (Khera et al., 1976) associated with tachycardia and first and second degree of heart blocks (Grabowski and Payne, 1980)

[a] As published.
[b] By the same route and at about the same stage of gestation as used in teratologic assessments.

Table 9-3. Category A: Known Human Teratogens (Specific Malformations at Apparently Nonmaternotoxic Dosages in Humans)

Compound	Salient malformation in humans	Animal species manifesting similar malformations
1. Diethylstilbestrol	Adenocarcinoma	Mouse, rat, monkey epithelial lesions of vagina
2–5. Ethisterone, 17-methyltestosterone, testosterone, and norethindrone	Masculinization of newborn female	Mouse, rat, guinea pig, monkey
6. Iodine, thiouracil	Goitrous cretinism	Mouse, rat, rabbit, guinea pig
7. Organic mercury	Infantile cerebral palsy	Cat, monkey
8. Tetracycline	Discoloration of bone and teeth	Rat, dog
9. Thalidomide	Phocomelia, amelia, and others	Nonhuman primates

Table 9-4. Category B: High Incidence of Malformations Induced in Animals by a Wide Maternotoxic Dose Range

Chemical	Species	Salient malformation	Reference
Copper citrate	Hamster	Cardiovascular defects	DiCarlo, 1980
Hydrocortisone	Hamster	Cleft palate	Shah and Chaudry, 1973
Adriamycin	Rat	Esophagel and intestinal atresia and tracheoesophagel fistula, cardiovascular and other defects	Thompson et al., 1978
Glycerol formal	Rat	Ventricular septal defect	Aliverti et al., 1980
Triamcinolone	Rat	Cleft palate	Walker, 1971
Vitamin A palmitate	Rat	Anophthalmia, cleft palate	Hayes et al., 1981

9.2.3. Category C

Category C consists of chemicals that are associated with reproductive toxicity occurring only at maternotoxic doses. The incidence of fetal malformations is low and that of runting is high, and the dose-response relationship for each individual malformation is generally absent. A specific pattern of fetal malformations, reported at maternotoxic doses of a large number of chemicals, has been tentatively related to maternal toxicity (Table 3-2, p. 15) as a possible causal factor (Khera, 1985). Approaches to defining the relationship of maternal and developmental toxicity have been considered by Schardein (1987). It is often possible to predict the threshold for fetal effects from the threshold level of overt toxicity in the maternal animals.

At maternotoxic doses, a chemical may produce fetal malformations, deaths, resorptions, or runting; yet there are chemicals that do not show any of these fetal effects. Also there are chemicals that cause only a few malformations characteristic of maternal-toxicity-associated pattern at the maternotoxic level. It is not known whether the absence and expression of a partial to complete pattern of developmental toxicity is related to a difference in type or severity of maternal toxicity. The possibility exists that a compound fetotoxic at maternotoxic doses in one species may be a selective teratogen in another species.

The different dose levels of some chemicals resulting in adverse fetal effects in different strains and species are not only generally indicative of relative fetal sensitivity but may also reflect the different doses inducing maternotoxic effects. The extreme susceptibility of maternal swine to the toxic action of vomitoxin paralleled the highest fetal sensitivity. The minimum fetotoxic and maternotoxic doses were 0.1 mg/kg in gilts (calculated from Friend et al., 1983), 1.8 mg/kg in rabbits (Khera et al., 1986), and 2 mg/kg in mice (Khera, 1984). These differences may partly depend on genetically determined factors, such as differences in enzyme levels or the ability to metabolize the parent compound and its degradation products.

In humans there are at least seven drugs (Table 9-5) that at overtly or potentially maternotoxic dose levels have been alleged to produce fetal wastage and a wide spectrum of congenital malformations involving multiple organ systems. Most of these drugs were associated with a low incidence of malformations. Because maternal toxicity data were insufficient, it is difficult to say whether these drugs belong to category B or C.

9.2.4. Category D

This category contains chemicals that produce neither embryo-fetal mortality nor fetal malformations in animals, even at maternotoxic doses. For humans

Table 9-5. Compounds Accepted as Human Teratogens That Show a Similar Pattern of Malformations Associated With Usually Maternotoxic Doses

Accutane
Aminopterin
Phenytoin
Trimethadione
Phenobarbital
Ethanol
Warfarin

a large number of compounds that have not been used at maternotoxic dosages and that do not seem to have a developmentally toxic potential at the usual therapeutic dose levels or at exposure levels can be tentatively placed in this category.

It has been reported that 41% of 165 compounds with no evidence of teratogencity in humans apparently produced adverse fetal effects in more than one animal species. Thus, it was concluded that animal data have limited value for making correlations of fetal responses between humans and experimental animals (Frankos, 1985). However, this conclusion should be qualified. A review of published data on 42 of these compounds that were positive in animal studies revealed marked differences in the dose levels involved as well as differences in the route and duration of dosing. In most instances doses in humans were at a therapeutic level, whereas doses designed to determine toxic effects in animals were much higher (Table 9-6). The comparison between the adverse fetal responses, obtained mostly at high doses in animals, and the no-effect level at low human exposure levels may not be valid. Excessively high test doses are known to alter maternal homeostasis by saturating metabolic pathways, altering osmolarity, and changing functions of organ systems as well as inducing pathological changes (e.g., cell death) in fetal tissue.

9.2.5. Chemicals of Ill-Defined Category

It is common to find chemicals that are difficult to classify because they do not fit neatly into any one category. There may be chemicals that are selective teratogens in one species but produce malformations only at maternotoxic doses in another species and may be nonteratogenic at maternotoxic doses in a third species. A conservative classification procedure is to use results from the species showing the selective action of the test chemical unless the metabolism of the chemical in this species is markedly different from that of humans. There would be chemicals difficult to categorize because they share properties characteristic of two categories or fall into different categories on testing in several species. Such chemicals, in the absence of metabolic data, would be moved from the lower to the higher category (i.e., from D to C, C to B, or B to A) for the purpose of assessing human safety. If the results of replicate studies of a chemical in the same species disagree because of a difference in test strains or for any other reason, then the validity of the protocol and the conduct of experiments would be the deciding factor.

For the purpose of extrapolation, the potential for human risk of chemicals in category A should be regarded as definite, B as probable, C as possible, and D as unlikely. It is understood, however, that a subsequent perception of real risk from a chemical may be at variance with this arbitrary estimation.

9.3. Methods of Risk Extrapolation

Extrapolation of human reproduction risks from animal data is complicated by the fact that the basic experimental unit is the complete maternal-fetal unit rather than the fetus only (Hogan and Hoel, 1982). In a more elaborate form, this is really a maternal-(preplacental or placental) membranes-embryo-fetus unit in which the morphology and functions of membranes are always changing in relation to the developing and differentiating embryo or fetus. Only fragmentary information is available on the functional role of the three components of the maternal-membranes-embryo-fetal complex in causing abnormal development. Factors that regulate phenomena such as maternal health and its influence on the embryo, distribution kinetics across fetal membranes, and concentration of chemicals at embryonal target cells in various experimental species have not been fully defined. The complexity and poor definition of these phenomena preclude the use of any standard extrapolation procedure.

The first step in human risk assessment is the evaluation of the type of adverse effects (fetal weight reduction, embryo-fetal mortality or malformations, fetal index, postnatal developmental organ function, etc.) and the doses at which these effects are observed in the species investigated. Then, from two or more test species, an estimate is made of the qualitative type of adverse effects expected and the safe exposure levels for humans. In the latter step, detailed consideration is given the overall toxicity data (WHO, 1984).

9.3.1. No-Effect Level Model and the Safety Factor

Traditionally, acceptable levels of exposure to compounds that produce toxic effects have been established by the application of a safety factor to the

Table 9-6. Compounds With No Positive Report of Teratogenicity in Humans: A Comparison of Human Usage Levels and Dosages Found Teratogenic in Animals

Compounds	Human exposure level, mg (or stated) per kg[a]	Minimum embryotoxic dose, mg (or as stated) per kg in mouse (m), hamster (h), rat (r), rabbit (rb) dose in animals	Approximate ratio, therapeutic dose/minimum embryotoxic dose
1. Acetazolamide	4	m 1,000, h 600, r 400	1:100
2. Actinomycin D	0.015	r 0.025–0.1	1:2
3. Aldrin	0.002 ug	m 25, h 50	1:12,000
4. L-Asparaginase	50–200 IU	m and r 100 IU, rb 500 IU	1:1
5. Caffeine	20–25	m 250, r 75–100, rb 100	1:3
6. Carbamazepine	24	r 250	1:10
7. Carbaryl	0.01	guinea pig 300, dog 6–50	1:30,000
8. Chloramphenicol	50–80	r 2,000–4,000	1:25
9. Chlorpropamide	2–5	r 200	1:100
10. Codeine	2.4	m 100	1:40
11. Colchicine	0.06–0.07	m 0.5–1.0, rb 0.1–0.5	1:1.5
12. Cyclizine	1–2	r 75–100	1:40
13. Dexamethasone	0.024 in nasal spray	m 6, r 4	1:150
14. Dieldrin	0.0001	m 3.4, r 3.4, h 30	1:> 3,000
15. Endrin	0.00001	m 2.5, r 0.58, h 5	1:> 3,000
16. 5-Fluorouracil	12	r 12–37	1:1
17. Griseofulvin	10–20	r 500–1,500	1:25
18. Haloperidol	0.04	m 14, h 80	1:350
19. Hydrocortisone	*b*	m 82.5, rb 10–50	
20. Hydroxyurea	20–30	r 137, cat 50, monkey 100	1:2.5
21. Meclizine	0.5–2	r10–40	1:20
22. Medroxyprogesterone acetate	0.1–0.2	rb 75, m 30	1:375
23. Mescaline metaproterenol		guinea pig 0.45–3.25	1:33

24. Methadone	0.1–0.3	m 24, h 67	1:80
25. Methylprednisolone	*b*	m 4–8	
26. Mirex		r 1–6	
27. Mitomycin C		m 5–10	
28. Morphine	1.2–3.6	m 400	1: > 100
29. Penicillin	up to 50,000 U	m 50,000–500,000 U	1:1
30. Perphenazine	0.3–0.6	r 20–150	1:200
31. Prednisolone	0.2–8	m 2.5, r 2.5, rb 0.1–0.25	80:1
32. Procarbazine	2–6	r 5–10, 10–40	1:1
33. Prochlorperazine	0.6–1.2	r 2.5–10	1:2
34. Pyrimethamine	0.5	h 160, r 12–40	1:24
35. Reserpine	0.001–0.025	r 100, rb 0.16, guinea pig 10	1:160
36. Sodium chloride	0.18	m 1,900	1: > 3,000
37. Sodium salicylate	72–108	m 500, r 250, cat 50, dog 400	2:1
38. Thioguanine	2	r 10	1:5
39. Triflouperazine	0.1–1.0	r 20–25	1:20
40. Urethane	0.012	r 1,000, m 15–1,500	1:1,000
41. Vinblastin	3.7 mg/m^2	m 2.5, r 0.25, h 0.25	
42. Vincristine	2 mg/m^2	h 0.1, m 0.15	

[a] Cited from McEvoy and McQuarrie, 1984; Goodman and Gilman, 1970; and other sources.
[b] Used for topical application.

no-observed-effect level (NOEL) in animal studies. This approach presupposes that the NOEL is equivalent to the actual no-effect level. Consideration of sample size not withstanding, the safety factor should take into account both intraspecies and interspecies differences in susceptibility to protect all segments of the population. The magnitude of the safety factor may also be influenced by the quality and scope of the available reproductive toxicity data. The safety-factor method meets a narrow definition of risk estimation and has inspired very little research in risk methodology other than generating data to determine intraspecies and interspecies variability in metabolism. Despite all these shortcomings, it is most widely used because of simplicity, practical usefulness, and the absence of a scientifically valid alternative.

Chemicals that have been adequately tested have revealed a dose-response relation. A similar relationship and, consequently, a threshold level is assumed for all chemicals yet to be tested. A NOEL, no-observed-adverse-effect level (NOAEL), and a lowest-observed-effect level (LOEL) are determined from animal data and, if available, human data. The NOEL, NOAEL, or LOEL is divided by the safety factor to determine an acceptable daily intake that is hypothetically harmless for human fetal development. The numerical value of this safety factor is influenced by the adequacy and extent of animal data, the intraspecies and interspecies variability of experimental animals, the shape of the dose-response curves, the imprecision of animal-to-human extrapolation, and the variability of the human population. Its value, therefore, is arbitrary because of the absence of research data directly relevant to the animal-to-human extrapolation process; to lay people its degree of magnitude is proportional to their degree of confidence in regulatory measures rather than to confidence in the quality and amount of data supporting the regulatory decisions.

If human exposure levels are close to clearly toxic effect levels in the progeny of a mammalian species, irrespective of the associated presence of maternal toxicity, continued human exposure is indefensible. This regulatory position should be violated for drugs only if benefits to maternal health outweigh fetal consequences, if the drug is intended for elderly and non-child-bearing individuals (Frederick, 1985), or if the test species is irrelevant to human risk because of the differences in maternal and fetal bioavailability, tissue distribution, or metabolism. With environmental pollutants, questions arise as to whether the existing exposure levels could present risks that are acceptable or unacceptable. The answer to these questions is in the domain of risk management and must be decided by considering economic and political factors in the given circumstances. The guidelines in this document deal with the estimation of human risk rather than with decision making to control risks.

A tentative safety factor should not be established from ambiguous and insufficient data unless there is a compelling need. Judging the suitability of data requires the opinion of scientists who understand the uncertainties of biological data. Similarly, the magnitude of the safety factor should be estimated only by individuals with in-depth knowledge of and experience in the field of reproductive toxicity.

The numerical value of a safety factor should reflect the type of fetal effect produced by a chemical in animal studies. It is not intended to consider any reproductive effect as being of less importance (i.e., fetal weight reduction, resorption, malformation, postnatal development, fertility, and organ function), and all effects are not of the same significance to the test species. The relevance of specific effects observed in a test species to potential specific effects in humans is unknown. A precise method is not currently available for quantifying the numerical value of a safety factor according to the relative significance of embryo-fetal effects in animal species and the probabilities of their occurrence in humans. In a survey of published data (Khera, 1984), the minimum dose reported to cause embryo-fetal effects in various animal species and the minimum dose (if not found, the minimum therapeutic dose) reported to cause congenital malformations or fetal wastage in humans were retrieved for a number of drugs. Of these, data on only 13 drugs were found sufficient for any comparison. The minimum dose for each species was divided by the minimum human dose, and the ratio was regarded as an index of animal species to human fetal sensitivity. The doses claimed as teratogenic for humans, in some cases, appeared so small that they may have been unrelated to the alleged effect. Conversely, in animal studies, the minimum doses causing adverse effects could actually have been lower than those reported because a complete teratogenic dose range for every drug was not investigated. Consequently, the crude estimates of animal fetus to human fetus sensitivity may be greater and, therefore, far more conservative than the real estimates. The accuracy of fetal sensitivity values were also limited by differences between human and animal studies in the route, level, and duration of exposure; stages of embryonic development exposed; criteria for evaluating fetal malformations; and distribution kinetics.

The ratio of animal to human fetal sensitivity ranged from 1 to 400 for six category-A drugs (thalidomide, methyltestosterone, ethisterone, northindrone, diethylstilbestrol, and tetracycline) investigated in seven mammalian species. The ratio ranged from 1 to 250 for seven category-C drugs (salicylates, phenobarbital, phenytoin, trimethadione, chlorpromazine, aminopterin, and methotrexate) studied in the same seven species (Khera, 1984). From these data the highest limits of 400 and 250 can be used as safety factors for human risk extrapolation of chemicals belonging to categories A and C, respectively, until better estimates are available. A safety factor with the value between 1:400 and 1:250 for chemicals in category C would be appropriate. For chemicals in category D, a safety factor ranging from 1 to 100 is probably valid. Another approach would be to use a base safety factor of 100, which is increased or decreased depending on the category to which a chemical is estimated to belong.

The above data indicate that the traditionally used 1,000-fold safety factor for category-A compounds is certainly adequate and may be conservative. The number of compounds for which data are available that fall in category A is very limited, and hence it would be advisable to maintain the 1,000-fold safety factor for this category.

9.3.2. Mathematical Models

Several methods are available for estimating exposure levels that would pose no potential risk to humans. Probit method of analysis of dose-response data for cleft palate induced by cortisone and 6-aminonicotinamide gave values that fit different linear and parallel curves (Biddle, 1977, 1978). The method is useful for comparing the reactivity (sensitivity) of strains and species. Its application for determining threshold levels is limited by the lack of sufficient evidence on the linearity of curves in the low-dose area. A pharmacokinetic model correlating maternal and fetal drug levels with teratogenic and fetotoxic responses (Young and Holson, 1978; Nau et al., 1981; Kimmel and Young, 1983) has been developed, but its use is restricted to those chemicals that are themselves proximate teratogens. This model fails to take into account maternotoxic responses and placental structure and activity that may adversely effect fetal development. Fabro et al. (1982) proposed a relative teratogenic index, which is a reflection of teratogenic hazard derived as a ratio of maternolethal dose (LD_{01}) to minimum teratogenic dose (tD_{05}). The concept, although meaningful for human risk assessment, is based on chemicals that are teratogenic only at maternotoxic doses, and it fails to make a distinction between malformation, resorption, and fetal death.

When behavioral and functional studies are considered in risk assessment, the relative toxicity in the adult and offspring must be considered, and subtle changes should not be given undue importance if present in the compromised offspring of compromised parents.

With the development of new areas of interest and technique, it has become possible to detect changes in additional organ systems. These new findings are sometimes prematurely considered to be useful for risk assessment. However, if a new test variable is considered appropriate for inclusion in a test battery, it should be validated, the results should be replicated, and the observed effects should be considered in terms of their relevance to humans.

Although most scientists are aware that "man is not a big rat" (Oser, 1981) and that rats have unique biological characteristics (Donaldson, 1924), the general public (Young, 1986) and particularly the legal profession seem to consider any adverse effect of a substance observed in animals to be directly related to possible adverse effects in humans. This invalid assumption is made frequently in terms of both cognitive and anatomical changes. Since the publication of results that related hyperactivity in children exposed in utero and/or during early postnatal life to CNS-active agents, the findings in animal models that showed hyperactivity similar to that in humans have been subject to general use and abuse (Hutchings, 1983).

In many instances the so-called subtle or even grossly apparent effects in offspring are frequently established using very sensitive methods, with relative disregard to the health of the mothers. The report of behavioral teratogens by Vorhees et al. (1979) is an excellent example of emphasis on evaluation of the offspring with minimal evaluation of the mothers.

Unequal emphasis on evaluating adult variables versus those of offspring has been common in the field and has not been restricted to postnatal behavioral and functional evaluations. For example, most of the developmental toxicity studies reported for benzene included relatively sensitive morphologic evaluations of the offspring (Coate et al., 1982, 1984; Green et al., 1978; Hazleton, 1982; Hudak and Ungvary, 1978; Hudak et al., 1980; Nawrot and Staples, 1979; Rodics et al., 1980; Tatrai et al., 1980a, 1980b). However, none of these studies considered the hematotoxic properties of the chemical in the maternal animal or sufficiently examined the well-documented endpoint of maternal body-weight gain. As a result, undue time and money was unnecessarily expended, and the chemical was inappropriately considered to possibly constitute a unique hazard to development. In fact, the dosages reported as teratogenic were those producing maternotoxic effects in other studies, and therefore were not uniquely hazardous to the conceptus.

Many researchers continue to report effects on developing fetuses or delivered offspring identified by ultrasensitive techniques, for example, by electron microscopic examination, without applying similarly sensitive techniques to evaluation of the maternal system. Thus, many publications appear to report undue risk to offspring in the absence of toxic effects in the mother, whereas these studies are actually basic research observations that are inappropriate for use in risk evaluation.

There is no question that basic research will continue to reveal new independent or interrelated subtle effects in offspring. From currently available information, however, no functional effect in offspring has been produced at a level that does not also result in classical endpoints of toxicity in maternal animals unless the agent has been identified as a developmental toxicant by classical developmental toxicity testing procedures (Johnson and Christian, 1986). Despite the justified interest in newer areas of research, at this time risk assessments and subsequent extrapolation should be made on the basis of validated methods and endpoints of developmental toxicity and maternal toxicity.

Current research into the basic biology of embryogenesis will continue to provide the basis for new developments in teratogen testing. Nonetheless, it is apparent that the testing methodologies currently available have limited application. Some of these tests are being used during product development and may find significant use in the area of product liability. In any case, as is true for genetic toxicology testing, in vitro teratogen assays may be best applied in a test battery scheme. The design and success of such test batteries is, however, largely dependent upon the development of a validation database. Approaches to the validation of in vitro methods in developmental toxicity testing have been suggested (Kimmel et al., 1982; Kimmel, 1985). The key element in validating these in vitro assay systems is their accuracy in predicting human teratogenicity. In this regard, the reference compounds of choice are the known human teratogens. However, the major teratogenesis database was derived from rodent studies (see Shepard, 1980); thus, from a statistical point of view, validation only by correlation would likely require the use of the rodent database. Uncertainty resulting from the use of correlation data of this type could be resolved if the in vitro test endpoint could be shown by a process of validation to be mechanistically allied with the processes of in situ teratogenic insult.

The question of mechanisms as it relates to in vitro test development is central to the process of validation and to test-battery design because there are probably multiple mechanisms involved in the induction of terata (Wilson, 1978). For example, assay systems that measure perturbations in chondrogenesis (e.g., limb-bud assays) may yield useful information concerning a test material's potential for inducing skeletal abnormalities but are less likely to identify teratogens or developmental toxins that act on embyronic tissues of nonmesenchymal origin.

The principal role of currently available cytotoxicity tests is to prescreen chemicals at an early stage in their development. Prescreening has the potential to distinguish those members of families of moities with similar "field" activity that are likely to pose developmentally toxic hazards. In vitro test development is, however, an emergent field that will continue to evolve. This evolution will accommodate the findings of basic research in developmental biology, which ideally will make mechanistic rather than correlative analyses possible.

The general scarcity of epidemiological data and the difficulties in controlling variables restrict the use of epidemiology in risk assessment. An increase in epidemiologic activity in the 1950s, particularly with the then virtually unknown tool of the case-control study, prompted many questions about the existing operations criteria for making causal inferences in a setting in which exposure cannot be controlled by the investigator. After much discussion, a set of seven general criteria that have been widely accepted and quoted was formalized by Hill (1971):

1. *Strength of association.* The stronger the observed association, the less likely that the finding is entirely due to sources of error that might distort the results.

2. *Dose-response effect.* The observation that frequency of disease increases with increasing exposure usually lends support to a cause-effect interpretation.
3. *Lack of temporal ambiguity.* It is important to show that the putative cause preceded the disease endpoint.
4. *Consistency of results.* If all the studies of a given association produce similar results, a causal interpretation is supported. This criterion is particularly important in an epidemiological context in which study populations, methods, and time periods are seldom truly replicated.
5. *Biological plausibility.* If the observed effect is congruent with current biological knowledge, a causal interpretation is more easily supported. However, this is a weak criterion in the sense that current knowledge might be inadequate to lend credence to the observations.
6. *Coherence.* If the observed effect is in agreement with other evidence regarding the disease (e.g., latent period or secular trend), a causal interpretation is strengthened. This criterion is somewhat redundant because it incorporates elements of consistency and plausibility.
7. *Specificity.* If the exposure is found to be associated with only one outcome, a causal interpretation is more plausible. This criterion was taken from Koch's postulates and applies better to infectious agents than to noncommunicable diseases.

These seven criteria can help to determine the likelihood of a causal relationship, but there are no firm rules for how they should be applied because no single criterion is either necessary or sufficient for a causal interpretation. Hill (1971) pointed out that a too strict adherence to any one of the criteria could result in incorrect conclusions.

Apart from the hypothesis-generating and risk assessment functions of descriptive and etiological epidemiological studies, covered in some detail in Chapter VIII, the same study designs have considerable potential but have been seldom used to complement the regulatory testing process for foods, drugs, and chemicals in the workplace.

Although toxicological testing strategies are being improved, many of the fundamental uncertainties that are associated with extrapolation of risk to humans are likely to remain. A logical extension of the regulatory process that would address this problem is the selected use of follow-up epidemiologic risk assessment studies. These studies would, within their inherent limitations, substantially reduce the uncertainty of whether or not an approved compound (e.g., a food additive, a prescription drug, or an industrial chemical) is actually creating a human health problem.

Monitoring and assessing human adverse-health effects from exposure to environmental agents of all types to confirm that regulatory decision making has been correct and that all control measures are effective are key aspects of the World Health Organization's Environmental Health Program (International Programme on Chemical Safety, WHO, 1985). It seems logical, therefore, to include epidemiogical follow-up data, although it will be necessary to set priorities (similar to those that have been proposed for mutagenicity testing [HWC,1986]) to determine which chemicals should be studied.

References

Aliverti, V., Bonanomi, L., Giavini, E., Leone, V.G., and Mariani, L. (1980) Effects of glycerol formal on embryonic development in the rat. *Toxicol. Appl. Pharmacol.* 56:93–100.

Andrew, F.D., and Staples, R.E. (1977) Prenatal toxicity of medroxyprogesterone acetate in rabbits, rats and mice. *Teratology* 15:25–32.

Biddle, F.G. (1977) 6-Aminonicotinamide-induced cleft palate in the mouse: The nature of the difference between the A/J and C57BL/6J strains in frequency of response and its genetic basis. *Teratology* 16:301–312.

Biddle, F.G. (1978) Use of dose-response relationships to discriminate between the mechanisms of cleft palate induction by different teratogens: An argument for discussion. *Teratology* 18:247–252.

Coate, W.B., Hoberman, A.M., and Durloo, R.S. (1982) Inhalation teratology study of benzene in rats. In *The Toxicology of Petroleum Hydrocarbons*, pp. 239–248, American Petroleum Institute, Washington, D.C.

Coate, W.B., Hoberman, A.M., and Durloo, R.S. (1984) Inhalation teratology study of benzene in rats. In *Applied Toxicology of Petroleum Hydrocarbons* (H.N. MacFarland, C.E. Holdsworth, J.A. MacGregor, R.W. Call, and M.L. Kane, Eds.), pp. 187–198, *Advances in Modern Environmental Toxicology*, Vol. 6, Princeton Scientific Publishers, Princeton, New Jersey.

DiCarlo, F.J., Jr. (1980) Syndromes of cardivascular malformations induced by copper citrate in hamsters. *Teratology* 21:89–101.

Donaldson, H.H. (1924) *The Rat–Data and Reference Tables for the Albino Rat (Mus norvegicus albinus) and the Norway Rat (Mus norvegicus)*, Ed. 2, Revised, Memoirs No. 6, Wistar Institute Anatomy and Biology, Philadelphia.

Fabro, S., Shull, G., and Brown, N.A. (1982) The relative teratogenic index and teratogenic potency: Proposed components of developmental toxicity risk assessment. *Teratogenesis Carcinog. Mutagen.* 2:61–76.

Frankos, V.H. (1985) FDA perspectives on the use of teratology data for human risk assessment. *Fundam. Appl. Toxicol.* 5:615–625.

Frederick, G.L. (1985) Assessment of teratogenic potential of therapeutic agents: Preclinical toxicology. *Clin. Invest. Med.* 8:323–327.

Friend, D.W., Trenholm, H.L., Fiser, P.S., Thompson, B.K., and Hartin, K.E. (1983) Effect on dam performance and fetal development of deoxynivalenol (vomitoxin) contaminated wheat in the diet of pregnant gilts. *Can. J. Anim. Sci.* 63:689–698.

Goodman, L.S., and Gilman, A. (1970) *The Pharmacological Basis of Therapeutics. A Textbook of Pharmacology, Toxicology, and Therapeutics for Physicians and Medical Students*, Ed. 4, MacMillian, New York.

Grabowski, C., and Payne, D.B. (1980) An electrocardiographic study of cardiovascular problems in Mirex-fed rat fetuses. *Teratology* 22:167–177.

Green, J.D., Leong, B.K.J., and Laskin, S. (1978) Inhaled benzene fetotoxicity in rats. *Toxicol. Appl. Pharmacol.* 46:9–18.

Hayes, W.C., Cobel-Geard, S.R., Hanley, T.R., Jr., Murray, J.S., Freshour, N.L., Rao, K.S., and John, J.A. (1981) Teratogenic effects of vitamin A palmitate in Fischer 344 rats. *Drug Chem. Toxicol.* 4:283–295.

Hazleton Laboratories America, Inc. (1982) Inhalation teratology study of benzene in rats (sponsored by the American Petroleum Institute, published as Coate et al. 1982, 1984).

Hill, A.B. (1971) *Principles of Medical Statistics*, Ed. 9, Oxford University Press, New York.

Hogan, M.D., and Hoel, D.G. (1982) Extrapolation to man. In *Principles and Methods of Toxicology* (A.W. Hayes, Ed.), pp. 711–731, Raven Press, New York.

Hudak, A., and Ungvary, G. (1978) Embryotoxic effects of benzene and its methyl derivatives: Toluene, xylene. *Toxicology* 11:55-63.

Hudak, A., Tatrai, E., Lorincz, M., Barcza, G., and Ungvary, G. (1980) Dependence on concentration of the embryotoxic action of benzene inhalation on CFY rats. *Morphol. Igazsagugyi Orv. Sz.* 20:261–268.

Hutchings, D.E. (1983) Behavioral teratology: A new frontier in neurobehavioral research. In *Teratogenesis and Reproductive Toxicology* (E.M. Johnson and D.M. Kochhar, Eds.), pp. 207–235, Springer-Verlag, Berlin.

HWC (National Health and Welfare and Environment, Canada) (1986) *Guidelines on the Use of Mutagenicity Tests in the Toxicology Evaluation of Chemicals, National Health and Welfare*, National Health and Welfare, Department of Environmental Contaminants Advisory Committee on Mutagenesis, Ottawa.

Johnson, E.M., and Christian, M.S. (1986) An overview of considerations basic to extrapolation, evaluation and interpretation of developmental toxicity safety evaluation data. Developmental Toxicology for the General Toxicologist (continuing education course), Society of Toxicology, Akron, Ohio.

Khera, K.S. (1973) N,N-Ethylenethiourea: Teratogenic study in rats and rabbits. *Teratology* 7:243–252.

Khera, K.S. (1984) Maternal toxicity—A possible factor in fetal malformations in mice. *Teratology* 29:411–416.

Khera, K.S. (1985) Maternal toxicity: A possible etiological factor in embryo-fetal deaths and fetal malformations of rabbit-rodent species. *Teratology* 31:129–153.

Khera, K.S., Arnold, D.L., Whalen, C., Angers, G., and Scott, P.M. (1984) Vomitoxin (4-deoxynivalenol): Effects on reproduction of mice and rats. *Toxicol. Appl. Pharmacol.* 74:345–356.

Khera, K.S., and Tryphonas, L. (1977) Ethylenethiourea-induced hydrocephalus: Pre- and postnatal pathogenesis in rats given a single oral dose during pregnancy. *Toxicol. Appl. Pharmacol.* 42:85–97.

Khera, K.S., Villeneuve, D.C., Terry, G., Panopio, L., Nash, L., and Trivett, G. (1976) Mirex: A teratogenicity, dominant lethal and tissue distribution study in rats. *Food Cosmet. Toxicol.* 14:25–29.

Khera, K.S., Whalen, C., and Angers, G. (1986) A teratology study on vomitoxin (4-deoxynivalenol) in rabbits. *Food Chem. Toxicol.* 24:421–424.

Kimmel, C.A., Buelke-Sam, J., and Adams, J. (1985) Collaborative Behavioral Teratology Study: Implications, current applications and future directions. *Neurobehav. Toxicol. Teratol.* 7:669–673.

Kimmel, C.A., and Young, J.F. (1983) Correlating pharmacokinetics and teratogenic endpoints. *Fundam. Appl. Toxicol.* 3:250–255.

Kimmel, G.L. (1981) Developmental aspects of chemical interaction with cellular receptors. In *Developmental Toxicology* (C. Kimmel and J. Buelke-Sam, Eds.), pp. 115–130, Raven Press, New York.

Kimmel, G.L. (1985) *In vitro* tests in screening teratogens: Considerations to aid the validation process. *Prog. Clin. Biol. Res.* 163C:259–263.

Kimmel, G.L., Smith, K., Kochhar, D.M., and Pratt, R.M. (1982) Overview of *in vitro* teratogenicity testing: Aspects of validation and application to screening. *Teratogenesis Carcinog. Mutagen.* 2:221–229.

Kury, G., Chaube, S., and Murphy, M.L. (1968) Teratogenic effects of some purine analogues on fetal rats. *Arch. Pathol.* 86:395–402.

Larsson, K.S., Elwin, C.E., Gabrielsson, J., Paalzow, L., and Wachmeister, C.A. (1982) Do teratogenicity tests serve their objectives? *Lancet* ii:439.

McEvoy, G.K., and McQuarrie, G.W. (1984) American Hospital Formulatory Service. American Society of Hospital Pharmacists, Bethesda, Maryland.

Murphy, M.L., Dagg, C.P., and Karnofsky, D.A. (1957) Comparison of teratogenic chemicals in the rat and chick embryos. *Pediatrics* 19:701–714.

Murphy, M.L., and Karnofsky, D.A. (1956) Effect of azaserine and other growth inhibiting agents on fetal development of the rat. *Cancer* 9:955–962.

Nau, H., Zierer, R., Spielmann, H., Neubert, D., and Gansau, C. (1981) A new model for embryotoxicity testing: Teratogenicity and pharmacokinetics of valproic acid following constant-rate administration in the mouse using human therapeutic drug and metabolite concentrations. *Life Sci.* 29:2803–2814.

Nawrot, P.S., and Staples, R.E. (1979) Embryo-fetal toxicity and teratogenicity of benzene and toluene in the mouse. *Teratology* 19:41A.

Oser, B.L. (1981) Man is not a big rat. Toxicology Forum Proceedings, Washington, D.C.

Rodics, K., Hudak, A., Tatrai, E., Lorincz, M., Barcza, G., and Ungvary, G. (1980) Embryotoxic effects of benzene inhalation. *Acta Morphol. Acad. Sci. Hung.* 28:220–221.

Schardein, J.L. (1976). *Drugs as Teratogens*, pp. 55–254, CRC Press, Cleveland, Ohio.

Schardein, J.L. (1987) Approaches to defining the relationship of maternal and developmental toxicity. *Teratogenesis Carcinog. Mutagen.* 7:255–271.

Shah, R.M., and Chaudhry, A.P. (1973) Hydrocortisone-induced cleft palate in hamsters. *Teratology* 7:191–194.

Shepard, T.H. (1980) *Catalog of Teratogenic Agents*, Ed. 3, Johns Hopkins University Press, Baltimore.

Staples, R.E. (1976) Predictiveness and limitations of test methods in teratology: Overview. *Environ. Health Perspect.* 18:95–96.

Strobino, B.R., Kline, J., and Stein, Z. (1978) Chemical and physical exposures of parents: Effects on human reproduction and offspring. *Early Hum. Dev.* 1:371–399.

Sunderman, F.W., Jr., Allpass, P.R., Mitchell, J.M., Baslet, R.C., and Albert, D.M. (1979) Eye malformations in rats: Induction by prenatal exposure to nickel carbonyl. *Science* 203:550–553.

Tatrai, E., Rodics, K., and Ungvary, G. (1980a) Embryotoxic effects of simultaneously applied exposure of benzene and toluene. *Folia Morphol. (Praha)* 28:286–289.

Tatrai, E., Ungvary, G., Hudak, A., Rodics, K., Lorincz, M., and Barcza, G. (1980b) Concentration dependence of the embryotoxic effects of benzene inhalation in CFY rats. *J. Hyg. Epidemiol. Microbiol. Immunol.* 24:363–371.

Thompson, D.J., Molello, J.A., and Strebing, R.J. (1978) Teratogenicity of adriamycin and daunomycin in the rat and rabbit. *Teratology* 17:151–157.

Vorhees, C.V., Brunner, R.L., and Butcher, R.E. (1979) Psychotropic drugs as behavioral teratogens. *Science* 205:1220–1225.

Walker, B.E. (1971) Induction of cleft palate in rats with antiinflammatory drugs. *Teratology* 4:39–42.

WHO (World Health Organization) (1984) *Principles for Evaluating Health Risks to Progeny Associated with Exposure to Chemicals during Pregnancy*, Environmental Health Criteria 30, World Health Organization, Geneva.

WHO (World Health Organization) (1985) *Guidelines for the Study of Genetic Effects in Human Populations*, International Programme on Chemical Safety, Environmental Health Criteria 46, World Health Organization, Geneva.

Wilson, J. (1978) *Handbook of Teratology*, Vol. 4, *Research Procedures and Data Analysis*, Plenum Press, New York.

Wilson, J.G. (1973) *Environment and Birth Defects*, pp.1–350, Academic Press, New York.

Young, B.B. (1986) Birth-defects testing should not ignore the mind. *Washington Post*, Health, p. 6, February 5, 1986.

Young, J.F., and Holson, J.F. (1978) Utility of pharmacokinetics in designing toxicological protocols and improving interspecies extrapolation. *J. Environ. Pathol. Toxicol.* 2:169–186.

Chapter X

Guidelines for Reproductive Toxicity Tests in Different Countries With a Proposal for International Harmonization

10.1. Introduction

Guidelines proposed by several countries for testing chemicals to determine their safety for human reproduction are essentially based on guidelines of the U.S. Food and Drug Administration (FDA) for pharmaceutical drugs (Goldenthal, 1966). The requirements for the type of tests to be conducted and the endpoints to be examined in these guidelines are shown in Table 10-1. Regulatory agencies of the United States, Europe, Canada, and Japan generally require studies on fertility and reproduction (Segment 1), embryotoxicity (including teratogenicity; Segment 2), and perinatal and postnatal effects (Segment 3) for safety assessment of pharmaceutical drugs. Embryotoxicity or teratogenicity studies (Segment 2) and multigeneration reproduction studies are required for agricultural and environmental chemicals and food additives.

The testing regimens for drugs and other chemicals (agricultural, environmental, etc.) reflect, in part, the difference in type of human exposure and, to some extent, differences in regulatory philosophy. Pharmaceutical drugs are intentionally given to humans, usually at relatively high doses and for limited periods of time. This allows the possibility of developmental toxicity during specific phases of reproduction, and drugs should therefore be examined in detail before approval for use. On the other hand, exposure to environmental chemicals is usually unintentional, occurs at very low doses, and may last for prolonged periods, even several generations. The tests for such chemicals tend to include a long-term exposure. Because accidental exposures to high

Table 10-1. Guidelines for Reproductive Toxicity Testing

Type of test	Endpoints to be examined
Fertility and general reproductive performance of males and females after continued treatment (Segment 1)	Signs of toxicity (effect on food intake, body weight, etc.), gonadal function. Libido, potency, fertility in males. Receptivity, fertility in females, ovulation, parturition, livebirths, litter size, survival, growth and development of offspring, sexual maturity.
Embryotoxicity (teratogenicity) after treatment of females only (Segment 2)	Signs of toxicity; preimplantation and postimplantation loss; live and dead fetuses; fetal weight; external, visceral and skeletal abnormalities; functional disorders.
Perinatal and postnatal study after treatment females only (Segment 3)	Signs of toxicity, time of onset and duration of parturition, fetal, growth development, lactation, survival and development of offspring.
Multigeneration study (two or more generations with one or more litters in each generation)	Fertility and general reproductive performance spanning several litters and generations. May include teratology study.

concentrations of chemicals do occur—for example, to methyl mercury in Japan and Iraq, to methyl isocyanate (an important component of pesticides) in Bhopal, India, and to dioxin in Seveso, Italy—more emphasis has been recently placed on embryotoxicity (teratogenicity) tests for environmental chemicals.

There is an important difference between drugs and environmental chemicals with respect to the detection of adverse effects on human reproduction. The total number of new drugs that reach the market each year is relatively limited, and they are released under strict government control and medical supervision. Thus, causal relationships between the drug and developmental toxicity are likely to be detected after the drug has been released and the exposure can be controlled. On the other hand, many thousands of industrial chemicals are produced each year and human exposure, which may be widespread but very low, is difficult or impossible to monitor. Epidemiological methods are unlikely to identify adverse effects of such low-level exposures unless specific suspicion is aroused.

The situation with pesticides and food additives lies somewhere between that of drugs and industrial chemicals because the number of new pesticides and food additives developed each year is small. Therefore, thorough animal testing is required before approval for use. Once these products are on the market, monitoring human exposure, as far as the general population is concerned, is usually impossible. However, useful data can be obtained by studies of individuals involved in manufacturing and spraying operations.

The governments of most leading countries have issued their own national guidelines for reproductive toxicity testing; moreover, different agencies in the same country have sometimes proposed different guidelines. As long as these differences are based on good scientific reasoning they are to be encouraged. Often, however, there is little scientific justification for the differences between guidelines. The resulting plethora of national guidelines of different countries may result in studies conducted according to one set of guidelines being unacceptable in another country. The disharmony in testing guidelines wastes money and animals, and the limited physical and personnel resources available to conduct such tests. It is understandably a great source of frustration to those responsible for meeting the various international requirements.

10.2. Current Guidelines for Reproductive Toxicity Testing

Although guidelines of most countries follow the same general design, there are many variations in the requirements. These involve the number of species and animals, duration of treatment, and types of examinations required. The variations proposed by different agencies in the United States, United Kingdom, Japan, EEC, WHO, and OECD have been described (ECETOC, 1983; Baeder et al., 1985). Except for Japan (MHW, 1984; Tanimura, 1985) and to some extent the Nordic countries (Nordic Council on Medicine, 1983), the formats used by the United States (Goldenthal, 1966), United Kingdom (Committee on Safety of Medicines, 1974), EEC (1983), and Canada (HWC, 1981) are generally very similar.

In the basic U.S. guidelines, the Segment 1 study involves treatment of the animals for a variable period of time before mating and throughout gestation and lactation (Table 10-2) and so includes the periods that are also covered in Segment 2 and 3 studies. The Japanese and Nordic guidelines divide the periods so that treatment periods for the Segments 1, 2, and 3 follow each other sequentially with no overlap.

The major deviations in Segment 1 lie in the Japanese requirements to discontinue dosing of females on day 6 (mouse) or 7 (rat) of pregnancy, to kill all test dams a day before term, and to count corpora lutea and live and dead (resorbing) fetuses. Though dosing is continued only until implantation (day

Table 10-2. Fertility and General Reproduction Study (Segment 1): Similarities and Differences Between Guidelines of United States (Goldenthal, 1966), EEC (1983), Canada (HWC, 1981), and Japan (MHW, 1984)

Protocol details	Similarities	Differences
Species	Rat	Japan—rat or mouse
Number of animals per group	At least 20 males and 20 females	Canada—at least 15 males and 30 females EEC—a minimum of 24 males and 24 females
Dosing: duration and frequency	Males: daily for a minimum of 60 days before mating Females: daily, initiated 14 days before mating	Canada—males for a minimum of 80 days before mating United States and Canada—females dosed until weaning EEC—females dosed throughout pregnancy Japan—females dosed until gestation day 6 (mouse) or 7 (rat)
Route of dosing	Intended human use	–
Number of treatment groups to a control group	Three	United States—at least two additional
Maximum mating period	Seven + seven days or two weeks	–
Examination for successful	Daily	–
Time of kill (F_0)	Males: after mating (at F_1 weaning)	Females: 1/2 on gestation day 13 (U.S.) and on day 20 (EEC), and remaining at weaning Japan—F_1 all, one day before term.
Records of F_0 data:		
(i) Body weight and food intake	Weekly	–
(ii) Water intake	Optional	–
(iii) Clinical signs and death	Daily	–
(iv) Mating, fertility and gestation indices	Calculated for intergroup differences	
(v) (a) Dams sacrificed: values on corpora lutea, implants, resporptions and live embryos	Calculated for intergroup differences in corpora lutea, total implants, live implants, and resorptions	Japan—(all dams sacrificed before term). Fetuses studied for sex and external malformations.

Table 10-2. (*Continued*)

Protocol details	Similarities	Differences
		EEC – fetuses examined for sex, weight, and anomalies (external, visceral, and skeletal) United States – embryos studied for viability.
(b) Dams allowed to deliver		
Data on F_1 progeny	(a) Lactation and viability index, body weight (postnatal days 1, 4, 12, and 21, and weekly until mating), food consumption (weekly from weaning until mating)	Japan – there is no F_1 progeny
	(b) Postnatal functional development – to be assessed	
	(c) Reproduction – evaluated using one male	United States – in special cases

5.5 in the mouse and day 6.5 in the rat), this approach is designed not only to study fertility but also to examine the development of fetuses. In contrast, the protocols of other countries prescribe killing one-half of the dams before term but allowing the remaining dams to litter normally. The assessment of effects on development is then made during prenatal and postnatal growth.

In the Segment 2 study the Japanese agencies require rats to be treated from days 7–17 of gestation, and approximately one-third of the dams are allowed to deliver and rear their progeny (Table 10-3). The progeny are examined for growth and development, functional and behavioral changes, and the ability to reproduce. Additionally, all pups delivered in a Segment 3 study also are reared and examined in ways similar to those in the Segment 2 study. All countries require morphologic assessment of fetuses in the embryotoxicity (teratogenicity) study, but the proportion of a litter allocated for skeletal and visceral investigations may vary. The differences in the design of perinatal and postnatal studies (Segment 3) among various countries (Table 10-4) are relatively minor and are not difficult to reconcile.

The British authorities (Committee on Safety of Medicines, 1974) recommend that the interim kill included in a Segment 1 study be performed on day

Table 10-3. Teratogenicity Study (Segment 2): Important Similarities and Differences Between Guidelines of United States, United Kingdom, EEC, Canada, OECD, and Japan

Protocol details	Similarities	Differences
Species	Rat (or mouse) and rabbit	
Number of animals per group	Rat 20; rabbit 12	Japan–(pharmaceuticals) per dose rodent 30
Duration of daily dosing	Rat, mouse 6–15; rabbit 6–18, may extend until one day prepartum	Japan–rat 7–17; mouse 6–15
Route	As in humans, or oral for chemicals	–
Number of treatment groups in addition to a control group	Three	United States (EPA) and OECD–one when negative at a single dose of 1 g/kg.
Body weight	Rodent–days 1, 6–16, and at sacrifice; rabbit–days 0, 6–18, 22, 26, and 29 of pregnancy	–
Food consumption	Pharmaceuticals, daily. Agricultural products and additives, weekly.	United Kingdom–agricultural products daily
Water consumption	–	–
Clinical signs and death	At least twice daily	–
Time of kill	Rat, day 20 of pregnancy; mouse, day 19 of pregnancy; rabbit, day 29 of pregnancy	Japan (pharmaceuticals)–2/3 on day 19 (mouse) or day 20 of pregnancy (rat), remaining allowed to litter. Number, viability, body weight, tests done for physical, functional, and reproductive development.
Corpora lutea, implantation, and resorption	Numbers to be determined	–
Placental weight	Not necessary	–
Fetuses: viable, dead, sex, weight and external anomalies	To be determined	–
Skeletal examination	Pharmaceuticals; rodent: 2/3 from each litter; rabbit: all fetuses	Japan–rodent: 1/2 from each litter
	Agricultural chemical and additives; rodent: 1/3 to 1/2 rats from each litter; rabbit: all fetuses	United Kingdom–rodent: 1/2 from each litter; rabbit: 2/3 from each litter

Table 10-3. (*Continued*)

Protocol details	Similarities	Differences
Visceral examination	Pharmaceuticals; rodent: 1/3 from each litter; rabbit: all fetuses Agricultural chemicals and additives; rodent: 1/2 to 2/3 from each litter; rabbit: all fetuses	Japan–rodent: 1/2 from each litter United Kingdom–rodent: 1/2 from each from each litter; rabbit 1/3 from each litter

Table 10-4. Perinatal and Postnatal Study (Segment 3): Important Similarities and Differences Between United States (Goldenthal, 1966), EEC (1983), Canada (HWC, 1981), and Japan (MHW, 1984)

Protocol detail	Similarities	Differences
Species	Rat	Japan–rat or mouse
Dosing: route, frequency, and duration	Intended human use, daily from day 16 of pregnancy until end of weaning F_1	Japan–pregnancy day 17 (rat) or 15 (mouse) until weaning
Treatment group in addition to a control group	Three	United States–at least two
Number of dams/group	20	EEC–at least 12
Time of sacrifice of F_0 dams	At weaning of F_1	–
Data on F_0 dams:		
(1) Body weight and food intake	Weekly	–
(2) Clinical signs and death	Daily	–
(3) Fertility and duration of gestation and lactation	Calculated for intergroup comparisons	
Data on F_1 progeny:		
(1) Number, sex, viability and lactation	Calculated for intergroup comparisons	Japan–if required each litter may be culled to eight offspring four males and four females) four days after birth
(2) Body weight	Recorded on postnatal days 1, 4, 12, and 21	Japan–weekly from weaning until mating
(3) Postnatal functional	Various functions assessed	United States–not assessed
(4) Reproductive functions	May be assessed	Japan–assessed on at least one male and one female from each litter

20 of gestation for an assessment of dominant lethal (mutagenic) effects in addition to the usual embryonic survival and teratogenic effects. The principle of the British guidelines has been adopted by EEC (1983). If the pharmacokinetics and metabolism in the Segment 1 study differ from those in Segment 2, teratological findings obtained in the two segments may be significantly different. In addition, killing some females on day 20 of gestation in Segment 1 allows study of both the effects on intrauterine growth and an assessment of survival in utero vs. that immediately postpartum. The British (EEC) protocol also recommends that the development of the offspring to maturity be monitored and the functional capacity of various systems, including reproduction, be assessed.

The original idea of the multigeneration study dates back to the 1930s with nutritional studies in rodents. The U.S. three-generation study was officially proposed for chemicals in food in 1949, and the protocols have been discussed in detail by Fitzhugh (1959, 1968). The broad outlines of a three-generation study (two litters per generation) with teratologic assessment included are given in Table 10-5. The salient differences in guidelines for multigeneration studies are presented in Table 10-6. Since that time a number of variants have appeared in the protocols in attempts to improve or simplify the design and conduct of such studies (EPA, 1978, 1979; FDA, 1970, 1982; NAS, 1977; MAFF, 1980, 1985). The major differences in the guidelines relate to period of treatment before mating, number of generations, number of litters per generation, and inclusion of teratology groups. All the guidelines provide protocols, the difference being whether the functional evaluation of the F_1 generation should include a study of its reproductive capacity—the production of an F_2 generation. EPA (1982, 1985) requires a two-generation reproduction study but (at the present time) does not request functional evaluation of organ systems other than those involved in reproduction.

As previously suggested, there may be a justification for the dissimilarity in the testing requirements for food additives or pesticides as opposed to those for pharmaceutical drugs. In the multigeneration study designed by the FDA (Fitzhugh, 1959), the protection from hazards is intended for individuals who may be unwittingly exposed to chemicals. Each generation may bear two litters; the second litter provides the parents of the next generation. In selected instances in the F_1 or F_2 generation, the parents are bred a third time to produce a litter that may be used for a teratogenicity study. Other than as outlined above, the remaining information collected is similar to the EPA two-generation studies (EPA, 1982, 1985), as described below. These studies are, in turn, essentially identical with the Japanese (MAFF, 1985) and OECD (1983) two-generation guidelines.

There do not appear to be any special requirements for food additives or new food materials that can justify a three-generation design. Because parent

Table 10-5. Broad Outlines of a Three-generation Study (Two Litters per Generation) with Teratologic Assessment Included

Number of generations	F_0, F_1, F_2, and F_3
Route of treatment	Via diet
Number of test groups in addition to a control group	Three
Number of animals per group	A minimum of 20 males and 20 pregnant females.
Mating procedure	Random, one female with one male for one week
Total mating period	Three weeks
Route of dosing	Oral, mixed in diet
Selection of offspring (F_1 or F_2) to produce subsequent generation	Random selection of 1–2 males or females from each litter
Effects on F_0 male and female parents	
Signs of toxicity and death	Daily observations
Body-weight and food-intake measurements	At weekly intervals until kill
Effects on offspring	
Signs of toxicity and death	Daily observation
Food-intake measurements	At weekly intervals
Body weight—F_1 and F_2	At birth; on postnatal days 1, 4, 14, and 21; and weekly afterwards
Body weight—F_3	At birth; on postnatal days 1, 4, and 21
Criteria for examination	Number of pups born and their survival at different times after birth, stillbirths, external abnormalities
Teratological assessment	Conducted by mating F_1 or F_2 generation
Gross necropsy	Conducted on each animal
Histopathology	High dose and control: target organs and all reproductive organs of F_0 and F_1 parent animals. If changes observed, examine lower levels.

(F_0) generation males are 5 weeks old when their 10-week treatment begins, and females are 13 weeks old as they begin their 2-week treatment before mating, both sexes will be sexually mature and capable of producing a normal litter if there is no test substance toxicity. Sexual maturity is attained in male

Table 10-6. Guidelines for Multigeneration Studies: Salient Differences[a]

Chemicals	Organization	Treatment period for F_0 parents, days	Requirements: generation, litters	Number of animals in each litter
Agricultural or general chemicals	U.S. Environmental Protection Agency (EPA, 1979)	males = 100 females = 100	$F_0 \rightarrow F_1 \rightarrow F_2$	males = 20 females = 20
	Health & Welfare Canada (1981)	males = spermatogenic cycle (54–70) females = two estrous cycles	$F_0 \nearrow F_{1a} \rightarrow F_{2b}$ (weaning) $F_{1a} \searrow F_{2b*}$ $F_0 \searrow F_{1b*}$	males = 20 females = 20
	Ministry of Agriculture and Forestry and Fisheries, Japan (1985 and proposed)	males = 56 females = 56	$F_0 \rightarrow F_1 \rightarrow F_2$ (weaning or sexual maturity)	males = 20 females = 20
	U.S. Organization for Economic Cooperation and Development (OECD, 1983)	males = 56 females = 56	$F_0 \rightarrow F_1 \rightarrow F_2$ (weaning)	males = 20 females = 20

Food additives	U.S. FDA (Food and Drug Administration, 1982)	males = 70 females = 14	$F_0 \to F_1 \to F_2$ (24h)	males = 20 females = 20
	Health & Welfare Canada**	males = 100 females = 14	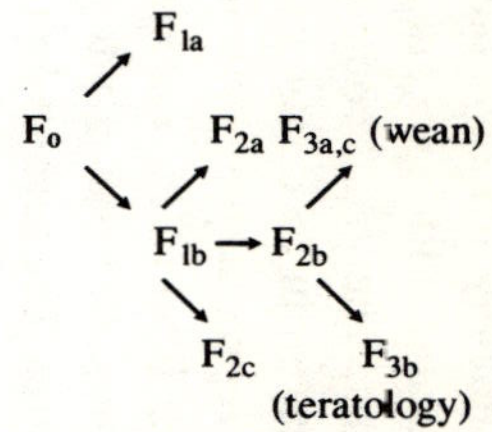	males = 20
	Ministry of Health and Welfare, Japan (guidelines in preparation)	males = 90 females = 90	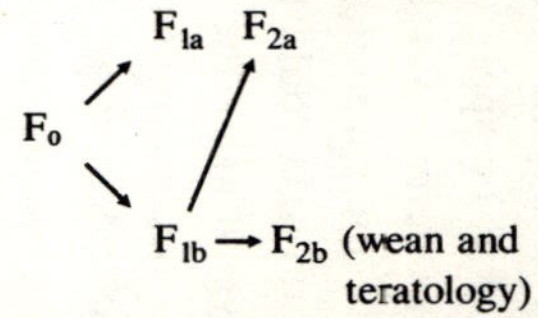	males = 20 females = 20

[a]Reprinted with permission from Tanimura, 1985, Pergamon Press plc.
* Required or desirable.
**Same as for agricultural chemicals.

Wistar rats at 75–100 days of age (Blazak et al., 1985) and Fischer-344 males at approximately 105 days of age (Blazak et al., 1985). Therefore, 15 weeks, or 105 days, would appear to be the age at which rats should produce litters of normal number and weight. These animals will be approximately 170 days old when they are mated to produce a second litter–still with a good chance of producing a normal litter. However, the same animals, bred a third time to produce an F_{2C} or F_{3C} generation, will be approximately 230 days old, and the appearance of latent spontaneous defects associated with age could seriously influence the results and their interpretations. The usual commercial feeds marketed for young growing animals could result in an unusual weight gain, a factor more critical than age in affecting reproductive performance (Palmer, 1981). There is no need to discard the first litter of each generation if they are sexually mature when bred because the animals necessary to establish the next generation may be chosen from this litter. The second (optional) litter in each generation could then be designated for a teratogenicity study, and the effect of confounding factors of age or weight on reproduction could be greatly reduced. The need for a third generation is warranted only if the reproductive performance in the second generation raises questions. Some of these issues were debated within the Food and Drug Administration 15 years ago but have remained unsettled (FDA, 1984).

The many variations that currently exist in a reproduction study design represent attempts to incorporate new ideas while designing a protocol covering every contingency. Although it is reasonable to expect protocols to reflect the advances in our knowledge, attempting to cover every eventuality of the reproductive process could be carried to an unreasonable extreme. The study design should be expected to accurately elicit unwanted effects, but the determination of the nature or mechanism of the effects should be reserved for separate, more specific studies. Thus, it is suggested that the best alternative at present to determine effects on reproduction and fertility incorporates most of the present EPA two-generation protocol (EPA, 1982, 1985), which is described below. It should be added that the OECD (1983) and Japanese (MAFF, 1985) two-generation reproduction studies are very similar to the EPA protocol and, therefore, will not be discussed, except to comment that a three-week mating period is unnecessary. Treatment of the F_0 generation males and females is initiated at approximately five weeks of age and of the F_1 generation (both sexes) at weaning. In each generation, treatment of the dams continues until weaning of their pups. If deemed necessary, each generation could be bred a second time, thus producing a second litter. Males are killed at the end of the first mating period unless a second litter is required. However, in some cases, males should be retained until an assessment of data from the first litter has been performed in order to determine if study of a second litter is needed. One male and one female are selected from the F_1 offspring from each litter and are crossmated between litters to produce an F_2 generation, thus avoiding brother-sister mating. The test substance may be given in the diet, in drinking water, by gavage, or by capsule. If the diet contains the test substance, it is possible for the offspring to consume a significant amount before weaning. The usual observations appropriate to a reproduction study are conducted, and a complete gross necropsy is performed on all animals. Target organs and organs of reproduction are preserved for histopathological examination. Complete histopathologic examination is conducted on reproductive organs of animals in the high-dose groups, which had been selected from the parent and F_1 generations for breeding. If a pathologic effect is demonstrated in the high-dose group, reproductive organs from the lower-dose groups and controls should be examined.

As will be seen in the following discussion, there is much in this EPA protocol that is unnecessary and some additional data that, if incorporated, would provide useful information. It is essential that the animals be at or near the peak of their reproductive capacity at the time of breeding; therefore, treatment should be initiated early enough to permit exposure to the test substance for the entire duration of a spermatogenesis cycle. A similar treatment period may not be necessary for females. Their dosing should begin when they are 13 weeks old and continue for 2 weeks before mating. This time is usually sufficient to initiate enzymatic induction or to modify metabolic processes. Thus, treated males and females both will be approximately 105–119 days of age at mating. It is, however, common practice to treat the females for the same duration of time as the males. A second litter (F_{1B}) could be produced by a parent generation approximately 170 days old at the time of mating, with fertility rate and litter size within the expected normal range if toxicity does not occur. To enhance the sensitivity of this second mating, it is suggested that mating of control and high-dose females should occur at the time of postpartum estrus because mating behavior, ovulation, implantation, and embryonic development are more easily disrupted by adverse environmental factors at this time than they are during a regular cycling estrus.

The uteri of females who failed to litter but were successfully mated (sperm in the vaginal smears) should be examined for the number and location of implantation sites and early resorptions (this presupposes that a second litter will not be required). A reduced litter size or pregnancy rate could be further evaluated if necessary by mating additional treated females with untreated males or vice versa, killing the females on day 13 of gestation, and examining their uteri for the number, location, and viability of the implantation sites. Appropriate functional evaluation of organ systems (for instance, renal clearance and lung capacity), defined by the knowledge acquired during toxicity testing of the compound, should be made on the offspring. Should an assessment of potential developmental toxicity be desired to examine reduced litter size or survival during the reproduction study, more females could be added in the high-dose group and killed on day 20–21 of gestation as is currently suggested by the British version of the FDA Segment 1.

There is usually no need for extensive histopathology in a protocol intended to screen for reproductive effects. However, a lack of effect on the incidence of pregnancy or litter size does not always mean that spermatogenesis was not altered. The effects of a test compound on male reproduction, according to the methods suggested in various guidelines, are difficult to separate from those on female reproduction because both males and females receive the test chemical during the premating period. Therefore, a microscopic examination of the testes should be conducted. It is suggested that the testes from the three test groups and the control group be appropriately fixed and that one testis of each male from the highest dose and control of each generation be examined routinely. This is to assure that testicular changes not detectable from fertility rate and litter size do not exist.

What are other reliable indicators of the existence of a toxic effect on the reproduction system of the rat? Evidence suggests that the testicular weight, unless dramatically decreased, is not a true measure of the rate of sperm production (Blazak et al., 1985). The number of sperm in the epididymis, being highly variable, has a questionable correlation with the rate of sperm production in the testis (Blazak et al., 1985). Evidence is accumulating that the rate of sperm production (SPR) and the characteristics of sperm movement could both be particularly sensitive measures of sperm damage – more so than histopathological examination of the testis. To determine these measurements, one testis and its cauda epididymis are homogenized separately. The homogenized testicular tissue is usually examined immediately to determine SPR, and the cauda epididymis is frozen for subsequent analysis of sperm movement. The details of these techniques are considered by Blazak et al. (1985). Although at present these techniques are not sufficiently validated to justify their routine use, continued publication of data derived from these techniques should be encouraged.

It bears repeating that although the present techniques for determining male rat reproductive toxicity are minimal, this is not reason enough to require routine histology of the entire male reproductive system nor the acceptance of new techniques, promising as they may seem, without careful evaluation of their value to routine screening in reproductive toxicology.

It is proposed that the above suggested modifications to the two-generation studies as required by both EPA (Federal Inseciticide, Fungicide and Rodenticide Act [FIFRA] and Toxic Substances Control Act [TSCA]) and OECD represent the current state of the art for rapid assessment of effects of test substances on reproduction of the rat. This proposal is based on the premise that these studies should be only designed only to demonstrate an effect on reproduction. If explanations of the observed effects are needed, specifically designed studies may be conducted later. It is not necessary or cost effective to create a second general toxicity study from a study intended to assess effects on reproduction.

10.3. Harmonization of Guidelines

It is generally agreed that efforts should be made towards harmonization of guidelines—and the feeling in virtually all scientific (as opposed to political) meetings is that this is desirable. Efforts should be made to organize meetings of experts in the various aspects of reproductive toxicology to draft guidelines that utilize the best points from the various guidelines to obtain the maximum amount of information without necessarily increasing the total amount of work or numbers of animals used. There might even be a reduction in the use of animals whenever possible by substituting alternative methods of study.

In the United States the Interagency Regulatory Liaison Group was established in an attempt to harmonize guidelines between the five regulatory agencies, but it has had limited success. The OECD has also developed guidelines for individual study designs that have had wide acceptance for industrial chemical toxicology, but it is still too early to assess their success.

Academic associations, such as the International Federation of Teratology Societies (IFTS), hold the greatest promise as centers for the harmonization of international guidelines. Scientists functioning cooperatively and collectively under the aegis of a professional society can impart more scientific objectivity to the task if they are unencumbered by organizational and political constraints. Strong personal biases tend to become mellowed, subdued, and more reasonable in the presence of eminent international scientific peers. A strong professional society has the potential to provide sound scientific advice for harmonization of guidelines that should be acceptable to most national regulatory agencies. Members of the IFTS might function under the sponsorship of the World Health Organization (WHO) in the harmonization of guidelines. Both the WHO (1984) and the IFTS (1986) have already held meetings relevant to this effort.

In the present international scene, the drugs, pesticides, industrial chemicals, food additives, and other substances used in individual countries, both developed and developing, have many more similarities than differences. Where special local factors operate, these must of course be taken into account, but in general the advantages to be gained by harmonization of guidelines far outweigh the disadvantages. Studies generated for one national regulatory authority would then be acceptable internationally. This would simplify the design, conduct, and evaluation of reproductive studies and improve the comparability of data between studies. It would prevent unnecessary duplication of work, reduce the time and cost for approval of new compounds, and reduce the numbers of animals used for research. If scientists agree that guidelines for testing can be modified and harmonized without any important loss of information relevant to human safety assessment, government administrators should be ready to act on such advice.

10.4. Steps Toward Harmonization

Attempts might be made by a group of scientists to define the essential components of individual tests and reconcile the differences between different protocols. It is recognized that variations would have to be made because of the specific characteristics of a test chemical or its use, but it is hoped that specific changes in individual protocols would be minimal and have sufficient scientific validity to be accepted internationally. Agreement would be sought on differences that exist in the design of the studies and on the endpoints to be assessed between different countries. Discussion and agreement are needed on the following issues:

1. Fertility and reproduction studies: species, age of animals at start of study, minimum treatment periods for males and females; whether some of the females should be examined during gestation and, if so, on what day of pregnancy; the types of tests to be carried out to assess the development of offspring.
2. Embryotoxicity and teratogenicity: species; whether some of the dams are allowed to deliver; how a litter should be divided for visceral and skeletal study; use of in vitro systems.
3. Perinatal and postnatal study: species, period of administration; types of examination of the offspring.
4. Multigeneration studies: species, ages of animals and durations of treatment before first mating; numbers of litters and generations; inclusion of teratology study groups.
5. Rationale for choice of doses, times, method, and duration of administration.
6. Role of pharmacokinetics in study design.
7. Design of behavioral and functional studies.

Clarification of these issues would lead to international standardization of guidelines. It would also provide for better interpretation of the results and more valid extrapolation of the finding to humans.

10.5. Conclusions

Sufficient knowledge has been acquired over the past 20 years of reproductive toxicity testing to allow for a more scientifically based discussion on the design and conduct of these studies. It is suggested that many of the variations in regulatory guidelines may not be of sufficient scientific importance to necessitate the conduct of several similar studies. The time seems right and the need is great for scientific organizations such as the International Federation of Teratology Societies in collaboration with the World Health Organization and other international bodies to attempt to achieve international agreement on harmonization of guidelines.

References

Baeder, C., Wickramaratne, G.A.S., Hummler, H., Merkle, J., Schon, H., and Tuchmann-Duplessis, H. (1985) Identification and assessment of the effects of chemicals on reproduction and development (reproductive toxicology). *Food Chem. Toxicol.* 23:377–388.

Blazak, W.F., Ernst, T.L., and Stewart, B.E. (1985) Potential indicators of reproductive toxicity: Testicular sperm production and epididymal sperm number, transit time, and motility in Fischer 344 rats. *Fundam. Appl. Toxicol.* 5:1097–1103.

Committee on the Safety of Medicines (1974) *Notes for Guidance on Reproduction Studies,* Department of Health and Social Security, Great Britain.

ECETOC (European Chemical Industry Ecology and Toxicology Center) (1983) *Identification and Assessment of the Effects of Chemicals on Reproduction and Development,* Monograph No. 5, European Chemical Industry Ecology and Toxicology Centre, Brussels, Belgium.

EEC (Commission of the European Communities) (1980) Council Recommendation of 26 October 1983 concerning tests relating to the placing on the market of proprietary medical products. *Offic. J. Eur. Comm.* L332:11–32.

EPA (Environmental Protection Agency) (1978) Proposed guidelines for registering pesticides. *Fed. Regist.* 43:37336–37403.

EPA (Environmental Protection Agency) (1979) Proposed health effects standards for Toxic Substances Control Act test rules. *Fed. Regist.* 44:44054–44094.

EPA (Environmental Protection Agency) (1982) Data requirements for pesticide registration. *Fed. Regist.* 49:42856–42905.

EPA (Environmental Protection Agency) (1985) Toxic Substances Control Act test guidelines. Reproductive and fertility effects; final rule. *Fed. Regist.* 50:39432–39433.

FDA (Food and Drug Administration Advisory Committee on Protocols for Safety Evaluations) (1970) Panel on Reproduction report on reproduction studies in the safety evaluation of food additives and pesticide residues. *Toxicol. Appl. Pharmacol.* 16:264–296.

FDA (Food and Drug Administration) (1982) *Toxicological Principles for the Safety Assessment of Direct Food Additives and Color Additives Used in Food, Guidelines for a Three-Generation Reproduction Toxicity Study With Optional Teratology Phase*, pp. 94–107, Washington, D.C.

FDA (Food and Drug Administration) (1984) *Review of Current Recommendations and Requirements for Reproductive and Developmental Toxicity Studies*, Food and Drug Administration Task Force on Reproductive and Developmental Toxicity, December 7, 1984.

Fitzhugh, O.G. (1959) *Appraisal of the Safety of Chemicals in Foods, Drugs, and Cosmetics*, Association of Food and Drug Officials of the United States, Topeka, Kansas.

Fitzhugh, O.G. (1968) Reproductive Tests. In *Modern Trends in Toxicology* (Boyland, E. and Goulding, R., Eds.), pp. 75–85, Butterworths, London.

Goldenthal, E.J. (1966) *Guidelines for Reproduction Studies for Safety Evaluation of Drugs for Human Use,* Drug Review Branch, Division of Toxicological Evaluation, Bureau of Science, Food and Drug Administration.

HWC (Health and Welfare Canada) (1981) *Preclinical Toxicologic Guidelines, 2–4, Reproduction Studies*, pp. 34–39, Health and Welfare Canada, Ottawa.

IFTS (International Federation of Teratology Societies) (1986) International Federation of Teratology Societies Symposium on International Standardization of Reproductive and Developmental Toxicity Testing. *Teratology* 33:37A–39A.

MAFF (Ministry of Agriculture, Forestry and Fisheries) (1980) *Guidelines for Studies on the Basis of Standards for the Evaluation of Feed Activities*, 98 pp, Bureau of Animal Industry, Ministry of Agriculture, Forestry and Fisheries, Tokyo, Japan.

MAFF (Ministry of Agriculture, Forestry and Fisheries) (1985) *Guidelines on Toxicology Study Data for Application of Agricultural Chemical Registration*, January 28, 1985, 59 Nohsan No. 4200, 55 pp, Ministry of Agriculture, Forestry and Fisheries, Tokyo, Japan.

MHW (Ministry of Health and Welfare) (1984) *Information on Guidelines of Toxicity Studies Required for Applications for Approval to Manufacture (Import) Drugs*. Notification No. 118, February 15, 1984, 26 pp., Pharmaceutical Affairs Bureau, Ministry of Health and Welfare, Tokyo, Japan.

NAS (National Academy of Sciences) (1977) *Principles and Procedures for Evaluating the Toxicity of Household Substances,* pp. 99–110, Publication 1138, National Research Council, Washington, D.C.

Nordic Council on Medicine (1983) *Drug Applications,* Nordic Guidelines, IIIC, Fetal toxicity, IIID, Fertility Studies, No. 12, pp. 30–31. NLN Publications, Uppsala, Sweden.

OECD (Organization for Economic Cooperation and Development) (1983) *Guidelines for Testing of Chemicals, Section 4—Health Effects*, No. 415 One-Generation Reproduction Toxicity Study, 8 pp., and No. 416 Two-Generation Reproduction Toxicity Study, 8 pp, Adopted May 26, 1983.

Palmer, A.K. (1981) Regulatory requirements for reproductive toxicology: Theory and practice. In *Developmental Toxicology* (C.A. Kimmel and J. Buelke-Sam, Eds.), pp. 259–287, Raven Press, New York.

Tanimura, T. (1985) Guidelines for developmental toxicity testing of chemicals in Japan. *Neurobehav. Toxicol. Teratol.* 7:647–652.

WHO (World Health Organization) (1984) *Principles for Evaluating Health Risks to Progeny Associated with Exposure to Chemicals during Pregnancy*, Environmental Health Criteria 30, World Health Organization, Geneva.